Acellus Learning Accelerator™

Courseware Development Guide for Educators

Dr. Roger E. Billings

IAS Press
Kansas City, Missouri

ISBN-13: 978-0-9631634-7-9

Published by International Academy of Science

11020 N Ambassador Drive
Kansas City, MO 64153

Version 1.0-20190815P

TABLE OF CONTENTS

ACKNOWLEDGMENTS

Special appreciation is expressed to the following wonderful people that have contributed much to make this book possible. My sincere thanks to each, and to so many others.

Dr. Maria Sanchez
Dr. Pajét Monet
Dr. Jacob Billings
Dr. Janice Carrell
Dr. Eileen Dayton
Marci Merkley
Ryan Etter

Chapter 1:
VISION OF ACELLUS

Learning Needs: As Diverse as Learners

All students – the ones who struggle and the ones who soar – deserve the opportunity to achieve their full potential.

We developed Acellus to apply science to the learning process, using advanced technology and feedback analysis to meet the needs of the diverse students schools need to serve. Schools that have adopted Acellus report reduced dropout rates, increased graduation rates, and growing success helping students transition to careers and college studies. Read their stories later in this chapter.

This book tells the story of Acellus – what it is, how it works, and how to use it to achieve desired outcomes.

Figure 1.1: Acellus speeds the learning process for every student.

What is Acellus?

Acellus is an interactive learning accelerator. It combines technology and learning science to help students learn more efficiently (in less time) and more effectively (greater mastery) (Figure 1.1).

Backed by scientific research, Acellus delivers online instruction, compliant with the latest standards, through high-definition video lessons made more engaging with multimedia and animation. Each lesson is carefully designed to connect with previously learned knowledge like interlocking building blocks. Students can log in to take lessons from school or home, on any device. The next time they log in, they're right where they left off.

Personalized instruction. Acellus adapts course delivery to the individual student using Intelligent Interaction (I^2) technology. After watching each video, students work problems that reinforce their understanding of that topic, for formative (during course) and summative (end-of-course) assessment. Students can advance to the next lesson as soon as they demonstrate mastery of the concept. When assessments reveal that a student is struggling, Acellus slows down and digs in, providing minor or major remediation and practice until the assessment shows the student is ready to move forward.

Less time spent on record keeping, more time for personalized attention. Acellus offloads much of the time-consuming work of grading so that teachers can devote more time to motivating students and helping them learn.

Teachers can closely monitor each student's progress from the Acellus Live Class Monitor (Figure 1.2). A quick glance shows which students are struggling and need more attention. In addition, Acellus Performance Reports help teachers identify students that are not succeeding and provide recommended intervention strategies. Teachers approve recommendations with a click of a button, and then Acellus immediately puts the plan into action.

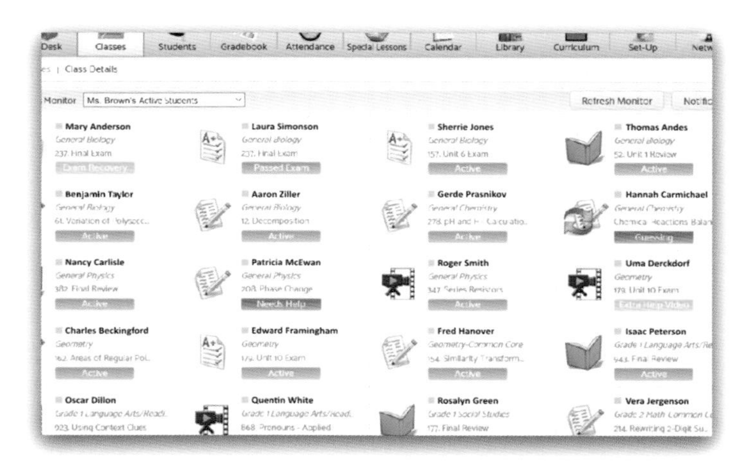

Figure 1.2: Acellus Live Class Monitor provides teachers with real-time information on the status of each student.

Continual improvement. The Acellus Learning System is designed to discover what does and does not work in the learning process. Continual refinements make course content more effective. Every night Acellus collects that day's student response data, which Acellus course developers study to better understand the learning process. The following night Acellus publishes new content that reflects insights from collected student responses. This continual feedback loop makes the learning experience continually more effective.

Real-World Results

After implementing Acellus, superintendents, school administrators, and teachers report significant gains in their standardized test results, accelerated student academic growth, and a notable decline in dropout rates.

Bridging the Gap in Student Comprehension

Ingleside Independent School District (ISD), Ingleside, Texas

"The special education students at Ingleside High School began by taking the Acellus Placement Exam to determine their independent levels in the subjects of English, Math, Science, and Social Studies (Figure 1.3).

"The amazing thing about the Acellus Program is that, through the Placement Exam, our students were placed into 13 different classes appropriate for their individual levels. This differentiation allows them to work at their own pace in completing the course work appropriate for their level. Consequently, our students have been able to make gains of 2 to 3 years in a single semester. The amount of curriculum covered thus far is astonishing. The "on task" and "task completion" behaviors have increased tremendously. All of this would not have been possible without the Acellus Learning System."[1] –Director of Special Education

Figure 1.3: Students at Ingleside ISD, Texas have made gains of 2-3 years in a single semester.

Higher Graduation Rates

Plainfield High School, Plainfield, New Jersey

"Since we've begun using Acellus, our graduation rate has steadily increased. Three years ago, our graduation rate was below 75%. This year our graduation rate is 90.1%. I can honestly say that a huge part of this success can be attributed to the Acellus program. Acellus has my highest endorsement."[2] —Principal

Blended Learning

St. Joseph High School, Metuchen, New Jersey

"Acellus has been an excellent resource for our students taking AP Calculus AB and BC – we use it as a supplement and as a standalone solution. This past year we had 23 out of 39 students pass the AP Exam with a perfect 5.0 score."[3] –Math Department Chair

---

[1] *Deborah Jones, Director of Special Education – Ingleside, TX*
[2] *Otis Brown, Principal – Plainfield High School, NJ*
[3] *Linda Muratore, Math Dept. Chair – St. Joseph High School, NJ*

Higher Standardized Assessment Scores, Leading to a Return of Students to the District

St. Louis Public Schools

"Since the implementation of Acellus in our Virtual School Program we have seen the following results:

- *Increase in the number of students graduating from our Virtual School Program using Acellus to recover lost credits. Initially there were 60 students which increased to 77 students the next year and 113 students the year after that.*
- *Increase in the number of students enrolled in the Virtual School Program from 201 students the first year to 388 students the second year and 458 students the third.*
- *The number of students enrolled as new or returning students using Acellus increased from 85 to 136 in the second and third years.*
- *The St. Louis Public Schools have saved $151,500 by implementing Acellus."* [4]

Improved Standardized Assessment Scores at the Middle School Level

Bertha Gilkey Pamoja Academy At Cole, St. Louis, MO

"The Bertha Gilkey Pamoja Academy @ Cole School used Acellus with their middle school students in science (Figure 1.4). This science class was taught by a substitute teacher. Below are the gains they made which are the results from the students Interim Assessment 1 given in September and Interim Assessment 2 given in November.

Figure 1.4: The goal of Acellus is not only to help students learn but also help them enjoy learning.

- *Gains were made this year with 7th & 8th grade students who used it last year as 6th & 7th grade students.*
- *The 7th graders went from 23% Below Basic to 4% Below Basic. They went from 8% Proficient to 16% Proficient.*
- *The 8th graders went from 15% Below Basic to 5% Below Basic. They went from 10% Proficient to 26% Proficient."* [5]

---

[4] *Carey Cunningham, Virtual Service Learning Coordinator – St. Louis Public Schools, MO*
[5] *Ibid.*

More Positive Attitudes Toward Learning

Chino Unified School District, California

> *"Some of these students who were struggling in Math, at home or in a school setting, and were frustrated and discouraged, once they went on Acellus, they were now encouraged. They were feeling they were getting better, they were learning, they were getting it. There was a light bulb that just said, "Hey I love math now!" When I started hearing that kind of feedback from my own students and the parents, I knew that Acellus was a great thing."*[6]

How to Put Acellus to Work in Your School

Choose a Deployment Model

Blended learning is growing in popularity.[7] The U.S. Department of Education reported that students with access to a combination of online and face-to-face instruction excel in relation to peers exposed to only one method of instruction.[8]

Acellus enables educators and administrators to adopt blended learning in a way that targets specific areas of need in their schools, using any combination of popular models.

Rotation model – Students rotate between online and face-to-face instruction.

Students rotate either on a fixed schedule or at the teacher's discretion. In some schools and courses students rotate among online learning, small-group instruction, and pencil-and-paper assignments at their desks. In others they rotate between online learning and some type of whole-group class discussion or project. In either case the clock or the teacher announces that the time has arrived to rotate, and students shift to their next assigned activity. The Rotation model includes four sub-models:

- Station rotation
- Lab rotation
- Flipped classroom
- Individual rotation

---

[6] *Sarah Sy, Instructor – Chino Unified School District, CA*

[7] *Jenny White, "Blended Learning Trendwatch: Models on the Rise," blendedlearning.org, September 27, 2018*

[8] *"Evaluation of Evidence-Based Practices in Online Learning: A Meta-Analysis and Review of Online Learning," U.S. Department of Education, 2010*

Flex model – An online course provides the backbone of student learning and is augmented by a face-to-face teacher.

The teacher of record is on site, and students learn mostly at a brick-and-mortar campus except when doing homework. Students move through a Flex course at their own pace, and the on-site teacher is on hand to offer help. In many programs the teacher initiates projects and discussions to enrich and deepen learning.

A la carte model – The course is taken entirely online to accompany a traditional brick-and-mortar program.

Students can take a la carte courses either on campus or off-site. Students take some courses a la carte and others face-to-face at the brick-and-mortar campus.

Enriched Virtual model – An online course that includes required face-to-face learning sessions.

Many enriched virtual programs begin as full-time online schools and then develop blended programs to provide students with brick-and-mortar school experiences. The enriched virtual model differs from the flipped classroom model because students typically do not meet face-to-face with their teachers every weekday. It differs from a fully online course in that face-to-face learning sessions are required; they are not optional or social events.

Figure 1.5: With Acellus Team Teach, teachers can bring Acellus into their classroom without a computer lab and create group collaboration.

No Computer Lab? Team Teach

Many teachers use Acellus to supplement direct instruction during the school day (Figure 1.5). Acellus Team Teach eliminates the need for a computer lab. Instead, teachers can select specific steps or concepts to show in class using a projector.

Teachers can select specific video lessons from any Acellus course to enhance the day's instruction and make it different. After presenting the videos, teachers can involve the class in group discussions on the topic, fostering collaboration between learners. Some educators use Team Teach in class and direct students to log into Acellus to complete their homework online. The corresponding lesson plans are already created and homework is graded automatically, giving teachers more time to focus on teaching.

Choose Programs

Schools around the U.S. are successfully using Acellus for programs ranging from Special Education and Credit Recovery to Gifted and Talented Education.

Special Education

The Acellus Special Education program is based on years of research by the International Academy of Science on how special needs students learn. The program applies proven methodologies to effectively reach these students so that they can quickly begin experiencing success, while preserving the rigor of standards-based courses.

Acellus Special Education courses are specifically designed for students with special needs, and include tools to assess, adapt, and individualize course material. Students can spend as much time as they need to approach or meet grade-level expectations.

Science, Technology, Engineering, and Mathematics (STEM)

Acellus STEM-10 prepares students for high-tech careers directly from high school (Figure 1.6). Unlike the usual games associated with STEM, STEM-10 is a cohesive 10-year program and a serious educational endeavor. Coding begins in the third grade and becomes more advanced each year. In the ninth grade, students branch into a career and technical field that matches their interest.

Coding instruction is delivered right through the Acellus system, making it suitable even in classrooms where a STEM-trained teacher is not available. The video-based coding courseware adapts to each student's individual needs, allowing schools to personalize the STEM instruction to each learner's level and skill set.

Figure 1.6: In the STEM-10 Level 2 coding course, students learn to program the AC-D2 robot - a robot that can walk, talk, dance, and sense distance.

Credit Recovery

When students start falling behind in their coursework or failing classes they need to graduate, timely intervention is imperative. Acellus provides an expansive selection of self-paced courses that students can complete on a flexible schedule. The courses help students bridge the gap between where they are academically and where they need to be in order to complete grade-level material.

Many schools that begin by using Acellus for Credit Recovery later shift to using it for remediation, to preempt failure. The result is a smaller Credit Recovery program – a good outcome.

Advanced Placement

Acellus offers an extensive selection of College Board-approved AP courses, allowing students to earn college credit while in high school. Acellus AP courses are rigorous while adaptive. The goal is to empower students not only to pass the test and acquire credit, but also to lay a firm foundation for future college studies.

Acellus AP courses are taught by certified AP instructors with years of experience teaching and preparing students for these rigorous exams. Acellus AP courses have been audited and approved by the College Board, making students eligible to receive college credit upon successful completion of the course and exam.

GED and Adult Education

Schools and adult learning centers can choose from an extensive selection of resources to provide academic and career training. To pass a high-school equivalency exam, students need a strong foundation in literacy, mathematics, social studies, and science — the four core areas tested by the GED, HiSET, and Test Assessing Secondary Completion (TASC). Acellus courses in each core area are specifically developed for each section of the exam. Students who lack the academic foundation to complete GED-level material can also take prerequisite courses through Acellus.

Career and Technical Education

In the influential article "The Silent Epidemic – Perspective of High School Dropouts," the authors reported that nearly half (47 percent) of people who had dropped out of school said a major reason was that "classes were not interesting."[9] These students

---

[9] *John M. Bridgeland, John J. DiIulio, Jr., Karen Burke Morison, "The Silent Epidemic – Perspective of High School Dropouts," Civic Enterprises, 2006*

did not see a link between the curriculum and the real world they would enter after high school.

Acellus CTE courses supplement or significantly expand CTE programs, combining academic anchor standards with the career and life-skills training essential for success in the real world (Figure 1.7). Students can take courses specific to their chosen career pathway. The course curricula adhere to state CTE guidelines while qualifying students to take industry-recognized certification exams. This opens the way for them to move into entry-level positions right out of high school.

Figure 1.7: Acellus Career and Technical Education (CTE) courses prepare students for careers in industry.

Independent Study

Schools need options for educating students outside of the brick-and-mortar school environment when students have health problems, are parents, work to support their families, or for various reasons cannot thrive in a regular classroom setting. Acellus gives students the opportunity to study from home – on their own schedule – without compromising academic rigor, standards, or quality.

Ensuring the academic integrity of a virtual independent study program can be challenging and time consuming for teachers and staff. Acellus automates much of the mundane work—logging every student login, answer, and grade, and generating reports – so that teachers can devote their time to educating students.

Intervention

With courses for academic as well as social and emotional learning, Acellus supports schools offering Response to Intervention (RTI) and Multi-Tiered System of Support (MTSS) programs. As a targeted intervention tool, Acellus links prescriptive assessments to customized content based on the student's skills and individual needs.

Using Prism Diagnostics® technology, Acellus identifies specific gaps in students' knowledge that make it difficult to learn a new concept. If gaps exist, the student receives additional instruction.

Figure 1.8: Dr. Pajét Monet addresses the issues of Social and Emotional Learning in the Acellus Emotional, Social, and Physical well-being courses.

Emotional, Social, and Physical Education

Today's educators are expected not only to engage students in learning but also to create an environment of positive thoughts and behaviors.

Research shows that social and emotional learning improves student achievement by an average of 11 percentile points, significantly improves student attitudes and behavior, and reduces depression and stress.[10] Physical fitness and healthy habits, in turn, have a long-lasting impact on students' happiness and well-being.

Acellus ESP (Emotional, Social, and Physical Education), helps students learn how to cultivate healthy relationships, have a caring attitude, and appreciate the importance of overall well-being and physical health (Figure 1.8). Courses are available for elementary, middle, and high school levels.

Summer School

Acellus offers diverse course offerings for summer school. For students who have fallen behind, schools can offer a complete credit recovery program. Basic courses needed for remediation help students complete the necessary grade-level courses and prepare them to pass end-of-course exams. For students on the other end of the spectrum, Acellus offers honors courses as well as a full suite of College Board-approved Advanced Placement courses.

After-School Programs

The pressure of preparing students for high-stakes testing in math and English Language Arts (ELA) leaves little time for enrichment activities such as STEM and robotics programs. Acellus After School programs provides help for struggling learners while giving other students an opportunity to engage in enrichment activities that may not fit into the regular school schedule.

---

[10] *Joseph Durlak, Roger P. Weissberg, Allison Dymnicki, Rebecca Taylor, & Kriston Schellinger, "The impact of enhancing students' social and emotional learning: a meta-analysis of school-based universal interventions," Child Development, January/February 2011, Volume 82, Number 1, Pages 405–432*

Teacher Development

The International Academy of Science (IAS) offers Master of Science (MSc) and Doctor of Science (ScD) degree programs for graduate students who want to specialize in technology-based instruction in a blended learning environment (Figure 1.9). These programs give graduate students hands-on experience applying both content and pedagogical knowledge in online and blended learning through the use of the Acellus Courseware Development System.

Figure 1.9: Graduate students of the International Academy of Science (IAS) specializing in technology-based instruction are able to study the learning process through the use of the Acellus Courseware Development System.

Gifted and Talented

High-achieving learners often need specialized instruction to realize their full potential. Providing a varied educational experience that meets these students' needs is an ongoing challenge that is difficult to address in a traditional classroom.

Acellus offers different levels of instruction so that students with advanced skill levels are not held back. An accelerated mode of Acellus, specifically developed for gifted students, features courses taught on a more advanced level than the traditional course, helping to keep gifted learners engaged.

Teachers choose the mode for each student. Acellus Success Zone reports suggest when to move students into the accelerated mode of a course.

English as a Second Language (ESL)

Approximately one in ten public school students is an English Language Learner (ELL).[11] As a result, schools across the country find themselves in the predicament of having to teach core academics for graduation without having the ability to verbally communicate with students. In some schools, the English Language Learners (ELL) program include native speakers of more than 50 languages.

The Acellus Discover English course teaches the English language and communication skills students need to thrive in school and the workforce. Developed for students in ESL Programs, Discover English is effective for students from any ethnic background. It uses the Universal Interaction Technique developed by Acellus to instruct students of various native languages simultaneously.

Discover English trains students on the syntax, vocabulary, and pronunciation they need to comprehend English in an everyday environment. This course is suited for students learning English for the first time as well as ESL students who need extra practice and help. Discover English can be used on its own or to supplement classroom teacher instruction in a blended learning environment.

Correctional Education

Hundreds of thousands of incarcerated adults and juveniles leave prisons and detention facilities each year. While many successfully transition back into their communities and become productive members of society, many others commit new crimes and are re-incarcerated.

According to a RAND study, "[i]nmates who participate in any kind of educational program behind bars—from remedial math to vocational auto shop to college-level courses—are up to 43 percent less likely to reoffend and return to prison."[12] Providing education to youths incarcerated in juvenile correctional facilities has been linked to lower recidivism rates and better reintegration rates when inmates leave detention. Education is a crucial service for correctional facilities. The two-fold challenge is providing education to a transient population with varying levels of mastery and doing it without Internet access.

---

[11] *Claudio Sanchez, "English Language Learners: How Your State Is Doing," NPR Ed, February 23, 2017*

[12] *"The Case for Correctional Education in U.S. Prisons," RAND, January 23, 2016*

Acellus has overcome these challenges by developing a media server that does not require a network connection. Inmates can access all Acellus educational lessons and complete their coursework without ever accessing the Internet.

Tutoring

Acellus is a cost-effective way to give students the extra help they need, when they need it. Struggling students benefit from tutoring – as do Honors students working hard to maintain their GPA for college admission.

Teachers can turn on the Acellus tutoring mode for any course. Then students can view individual video lessons on any concept taught in the course to deepen their understanding (Figure 1.10).

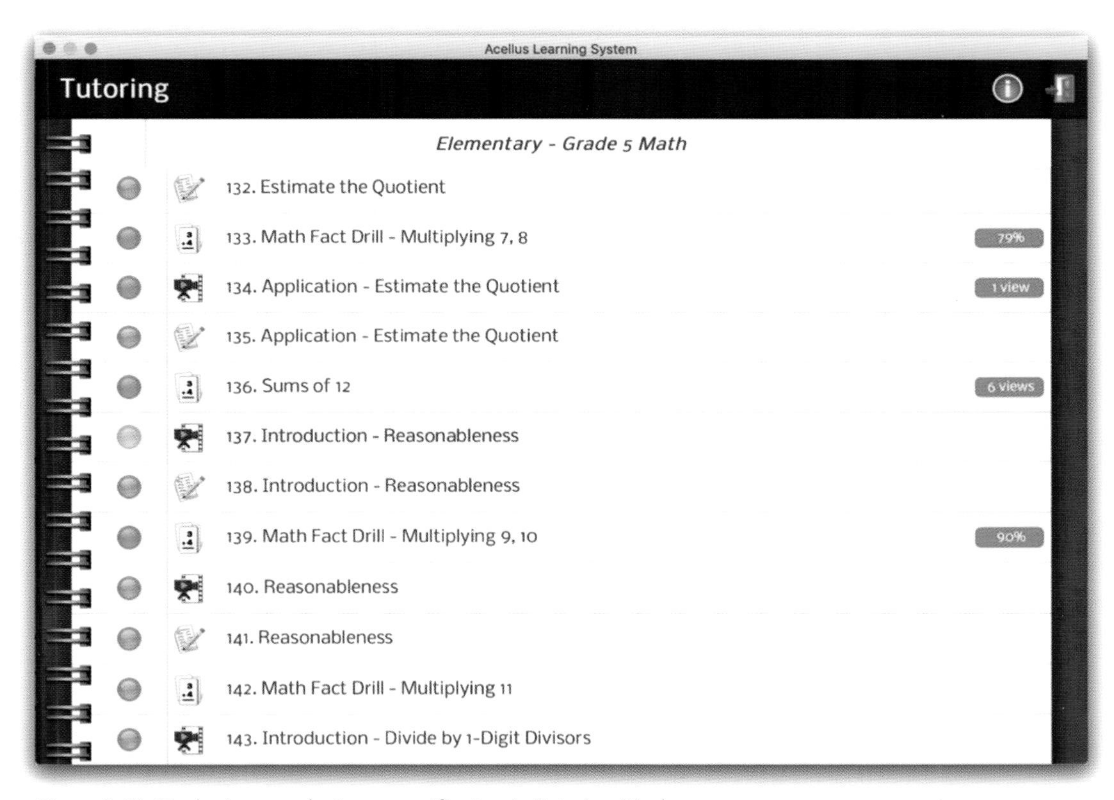

Figure 1.10: Students can select any specific step in Tutoring Mode.

College Prep

Acellus offers several courses for students preparing for college:

- *Investigating Careers* provides an overview of various careers – many of which students may not be familiar with – along with the education, training, and skills required for each.
- *College and Career Readiness* imparts the basic knowledge and skills students need to set attainable goals leading them onto a path of success.
- Specialized exam prep courses for college entrance and AP exams familiarize students with test question formats and provide time-management tips for test day. Knowing what to expect on a high-stakes test is almost as important as knowing the material covered.

Chapter 2:

THE SCIENCE BEHIND THE ACELLUS LEARNING ACCELERATOR™

So what is the "secret sauce" that makes Acellus work like magic? It's science. Building on modern research, Acellus developers study the way students learn, identify obstacles to learning, and experiment with different techniques rooted in cognitive science to discover the ones that accelerate learning. This chapter describes the science behind the Acellus Learning Accelerator.

Continual Feedback

Acellus collects feedback to continually refine the learning process. The feedback process occurs every night, when the current day's student response data is uploaded to the Acellus system for analysis by researchers at the International Academy of Science (Figure 2.1). Here's the process:

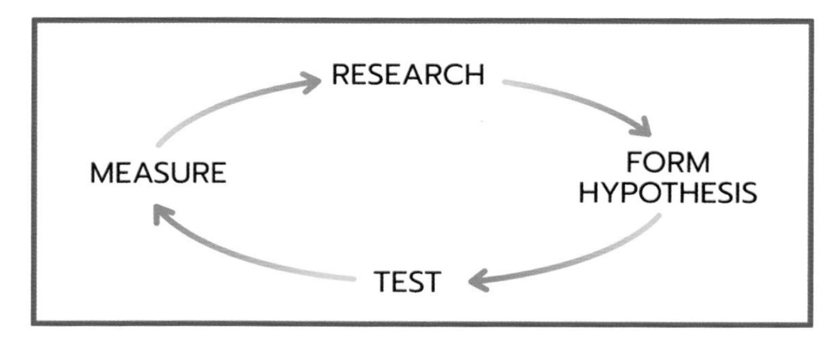

Figure 2.1: Acellus incorporates the scientific method and feedback process.

- Researchers study student responses to find areas where many students have difficulty.
- They form a hypothesis to explain why the concept is particularly challenging.
- They modify course content to test the hypothesis.
- Revised content is released to Acellus servers each evening.
- The Acellus system compiles student response data for the new problems to measure whether student understanding has improved.
- If student understanding did not improve, researchers form a new hypothesis and repeat the feedback cycle.

Through the scientific method and the power of feedback, Acellus researchers continually improve their understanding of how students learn, making the learning process more relevant, engaging, and effective.

The Success Zone Philosophy

Students who are not succeeding become discouraged. Some may feel like failures or lose hope that they will ever move forward and succeed. While teachers on the front line know how to motivate students, they don't have the time to personalize instruction for each student's needs – those who struggle and those who excel.

To address the factors inhibiting student success, Acellus researchers developed the Acellus Success Zone model. The premise is that 100% of students can succeed, where success is defined as making forward progress in the course with an average score of 70% or better. Success Zone gives teachers the data and intervention options to help students move forward. The outcomes are improved benchmarks such as attendance, course completion, diploma readiness, and college preparedness.

Here's how the Success Zone works:

- Students start the Acellus course in Normal Mode – the pace and complexity appropriate for the majority of students. The Acellus system continually monitors each student's progress and performance, flagging students unable to achieve at least 70% on lessons and exams. These students are potentially wasting their time struggling – not learning.
- When a student struggles in the course, Acellus recommends moving the student to the Slow Mode. This version of the course presents lesson content at a slower pace and might include videos with more basic content. Students doing exceptionally well in a course, on the other hand, while still being considered in their Success Zone, may benefit from an extra challenge. Acellus recommends moving these gifted students to the Accelerated Mode, which presents more comprehensive lessons and additional, more rigorous problems.
- Recommendations are compiled monthly into an online Performance Report that teachers can view at any time (Figure 2.2). The report shows the percentage of students in the class working in their Success Zone, which students are outside their Success Zones, overall attendance, and average class scores. The report provides specific recommendations on how to help struggling students and increase class scores – for example, encouraging students with poor attendance to consistently log in and complete their online coursework or moving students to Slow Mode or Accelerated Mode. Teachers approve the recommendations at their discretion and changes take effect right away.
- Teachers are given a rating that reflects the percentage of their students working in their Success Zone. As more students in a class move into their Success Zone, teacher ratings improve.

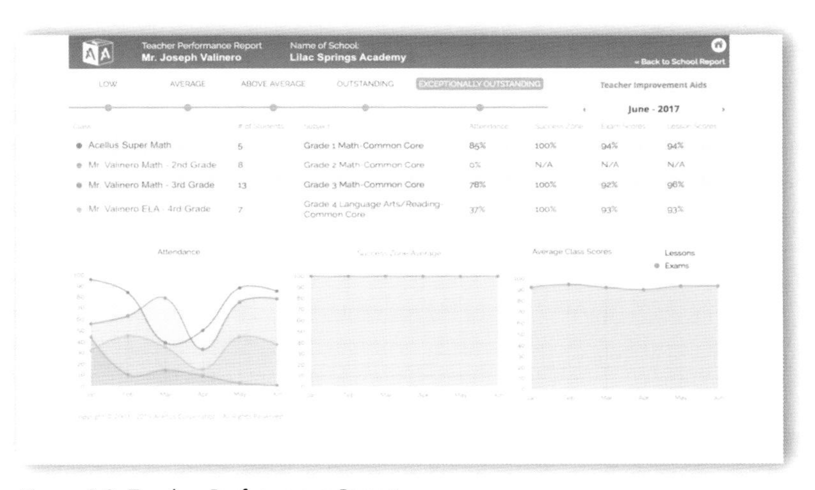

Figure 2.2: Teacher Performance Report

Principals can view the School Performance Report to quickly see how effective different teachers are at helping students move into their Success Zones (Figure 2.3). The report shows the school's average attendance, Success Zone average, and lesson and exam averages. Principals can drill down deeper to see classes with low attendance or a high percentage of their students working outside of their Success Zone. School administrators and board members can view District-wide Performance Reports to compare schools. All reports are updated monthly and show historical trends and progress.

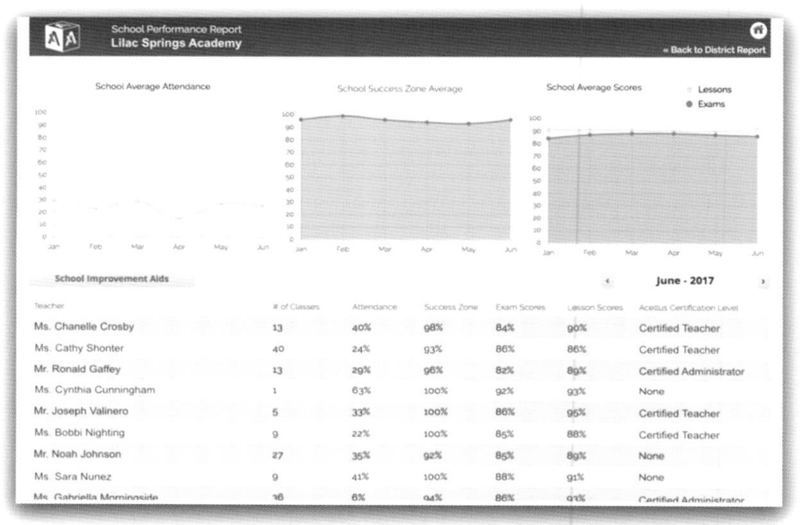

Teacher	# of Classes	Attendance	Success Zone	Exam Scores	Lesson Scores	Acellus Certification Level
Ms. Chanelle Crosby	13	40%	98%	84%	90%	Certified Teacher
Ms. Cathy Shonter	40	24%	93%	86%	86%	Certified Teacher
Mr. Ronald Gaffey	13	29%	96%	82%	89%	Certified Administrator
Ms. Cynthia Cunningham	1	63%	100%	92%	93%	None
Mr. Joseph Valinero	5	33%	100%	86%	95%	Certified Teacher
Ms. Bobbi Nighting	9	22%	100%	85%	88%	Certified Teacher
Mr. Noah Johnson	27	35%	92%	85%	89%	None
Ms. Sara Nunez	9	41%	100%	88%	91%	None
Ms. Gabriella Morningside	36	6%	94%	86%	93%	Certified Administrator

Figure 2.3: School Performance Report

Prism Diagnostics®

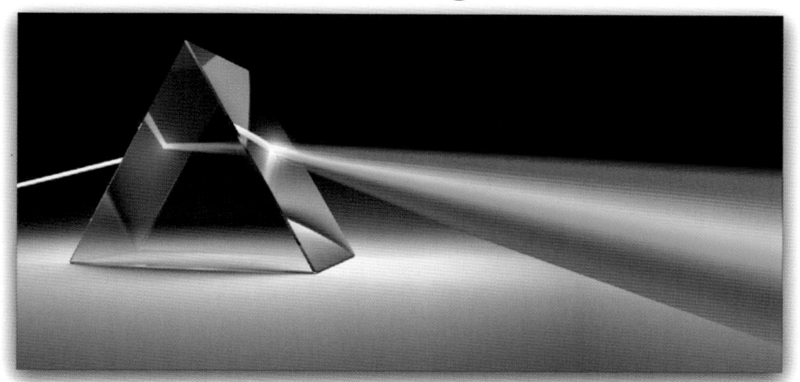

Figure 2.4: Prism Diagnostics separates students into groups based on similar deficiencies.

Prism Diagnostics (Figure 2.4) is made up of multiple modes or categories, each one designed to detect a specific problem or condition regarding student learning, thereby enabling the appropriate action or response. Here are the diagnostic modes and how they are each detected by the system:

"Stuck" Mode

After completing a video-based lesson, the student gets "stuck" when trying to work the assessment problems for the lesson. In Acellus, "stuck" is never good and not acceptable. "Stuck" students quickly become discouraged and negative.

Whenever a student is asked to work a problem as part of a lesson, a Help flag is provided on the same screen as the problem. Students learn to reach out for help in this manner, knowing that doing so will not negatively impact their scores. The system responds with Help Videos, which will be discussed in the next section.

Wrong Answer Analysis

Every time a student enters a response to an assessment question, Acellus checks the answer to see if it is correct. If the answer is not correct, the system searches the database to see if the incorrect answer entered is one of the "special" wrong answers recognized by the system. Sometimes, specific holes in students' understanding of a concept can be identified by analyzing their answers.

As a simple example, consider an assessment question asking students how many days are in a week. If the students answer "7 days," they have the correct answer and have demonstrated that they understand the concept of this lesson. If they give the wrong answer of "30 days," they are confusing weeks with months. If they answer "365 days," they are confusing weeks with years. In either case, they are demonstrating a need for narrowly-targeted instruction.

Wrong Answer Analysis is a powerful learning deficiency diagnostic which, when used properly, provides the ability to give struggling students the exact help required precisely when it is needed.

Progress Analysis

The Success Zone mentioned earlier is an Acellus tool used to motivate teachers and to keep the focus on student success and progress. It is also a powerful tool to identify when students have a chronic need for help.

Sometimes a simple fix is not enough. In these cases, Acellus provides each student with in-depth instruction referred to as the Recovery Mode. Student Progress Analysis consists of a complex set of algorithms which include rate of progress, overall scores on lessons and exams, and other components as applicable. All of these parameters are analyzed on a periodic basis to identify when a student would benefit from the Recovery Mode.

Vectored Instruction

When Acellus, using Prism Diagnostics, detects a deficiency, it instigates specialized instruction targeting the specific deficiency. This process is referred to as Vectored Instruction. It operates differently in response to each of the diagnostic modes.

"Stuck" Mode — Special Help Videos

In the "stuck" diagnostic mode, students are unable to work assessment problems, so they reach out for help by selecting the Help button. At this moment, the system knows exactly which problem the student is working on and is therefore able to instantly pull up a Help Video lesson that covers the same problem the student is trying to work but with different values or parameters. Watching the teacher work the problem on the screen helps the student see how to proceed but does not give the answer. After watching the Help Video, the student is again challenged to answer the original problem.

The Acellus system monitors the frequency of students needing Help Videos. Using this data, the courseware development team evaluates enhancements they could make to the main lesson video that would better prepare students to answer assessment problems. They also evaluate the problems themselves to see if they could be made more clear. On lessons teaching difficult concepts, special "teaching problems" are added to the lesson which are given to students before the assessment problems and often are effective in preparing students to handle the full-fledged assessments that follow.

The philosophy of Acellus is to provide immediate support, seeking to empower students to be able to work problems rather than just giving them the answers. In addition, lesson and exam problems are provided to students randomly so that students do not all get the same set of problems.

Wrong Answer Analysis — Deficiency Recovery Videos

Wrong Answer Analysis is one of the most powerful and effective tools in Acellus. At the very moment a student enters a wrong answer, the system responds with a deficiency recovery video targeting the specific flaw it discovered by analyzing the wrong answer.

Many times the reason students get discouraged about learning is that they keep getting things wrong and they cannot figure out why. They finally give up, thinking they are just "stupid." This is a tragic conclusion, leading to other failures and problems. Often, the deficiency is the result of some simple concept they have missed, often years before, but which is now needed to learn the new material of the present lesson.

When the deficiency recovery video feature of Acellus went live, school districts reported a measurable difference in the attitudes of students using the system, including improved scores on standardized tests and lowered dropout rates.

Progress Analysis — Vectored Instruction

The student Progress Analysis tool identifies students with chronic deficiencies. These students are placed into the Recovery Modes to help them get back on track.

From trial and error, researchers observed that when students with chronic deficiencies were placed back into classes below grade level, they began making more progress in their learning. However, this created problems since students often needed to be put back multiple grade levels before they were able to handle the lesson material. Putting students below their grade level was discouraging to some students and created problems for school districts required by law to have students in grade-level classes.

The solution was Vectored Instruction – the best of both worlds. It was born on the realization that even when students are behind, they only need specific lessons from past courses. A special database was created for each lesson of a course which listed the concepts taught in the lesson and identified the lessons from prior courses which taught concepts a student would need to be able to master the new material.

Vectored Instruction builds on the "mental models" theory articulated by Kenneth Craik in his 1943 book *The Nature of Explanation*. According to Craik, mental models are an individual's own view of reality. A young student's mental model of a dog, for instance, might be that it's furry, comes in a variety of colors and sizes, likes to fetch, and sometimes bites. Children's simplistic mental models are often riddled with misconceptions and falsehoods. As the child grows and matures, these models become more complex – and, hopefully, more aligned with reality.

The smallest learning component is a *knowledge bit* (Figure 2.5). The dog attributes mentioned above are knowledge bits. *Concepts* are groups of knowledge bits that provide meaning when they come together. Until there's a concept, it is difficult to know what to do with knowledge bits – or to retain them. Remembering the string l-e-

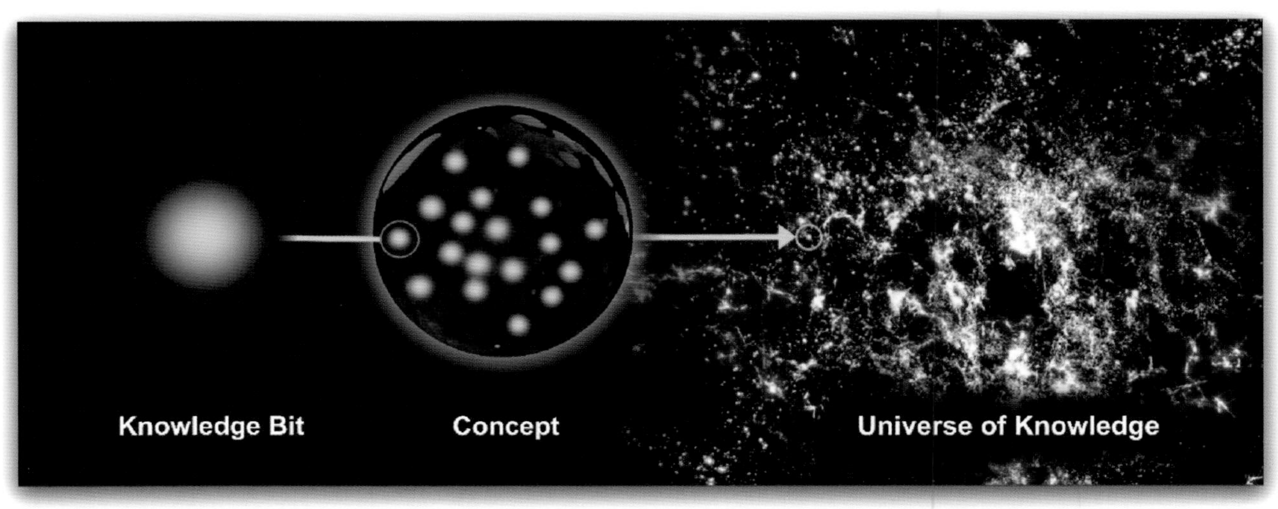

Figure 2.5: Interlinked knowledge bits form concepts; interlinked concepts form a person's universe of knowledge.

p-i-n-c is more difficult than p-e-n-c-i-l because the latter is part of a concept. Remembering the string 0-1-1-2-3-5-8-13-21 is easier if the learner understands that it is a Fibonacci Sequence describing the spiral shape in nature, where each number is the sum of the two preceding numbers.

Figure 2.6: Curriculum directors assemble need-to-know concepts to formulate a curriculum.

Education helps students accumulate bits of knowledge and relate them to a concept. As learning continues, the concept becomes more sophisticated and begins to interconnect with other concepts.

Interconnected concepts form a master conscious model – the universe of knowledge shown in Figure 2.5. The more students learn, the more complex their universe of knowledge becomes.

Curricula contain the need-to-know concepts identified by curriculum development teams. It's the job of educators to make sure that students master these concepts (Figure 2.6).

In Acellus, each lesson conveys a single concept. Students who are stuck are presented with remediation, right when the problem is fresh on the student's mind. The remediation is presented with the right magnitude and direction (like a vector) to fill the hole in understanding and put the student back on course.

Figure 2.7: Acellus Vectored Instruction

Vectored Instruction, illustrated in Figure 2.7, works like this:

- Researchers analyze student response data and diagnose deficiencies.
- Researchers analyze the concept of a lesson and determine which previously learned foundational concepts are required to master this new material.
- Acellus provides laser-precision instruction targeting the student's unique deficiency.

When a student has a small defect in understanding, deficiency recovery videos provide targeted instruction to fix a minute flaw, empowering the student to move forward again (Figure 2.8). For small deficiencies, this is all that is needed.

Figure 2.8: Deficiency recovery videos help correct minute flaws.

Other times, however, students have a chronic problem. They are not sufficiently prepared for the course that they are in. Vectored Instruction focuses on solving this more challenging problem. Vectored Instruction, when needed, is deployed in three levels:

- When Acellus diagnoses a persistent deficiency, students first go into Recovery Mode, where they receive more basic instruction and practice on the concept

Figure 2.9: Students with a persistent deficiency go into Recovery Mode.

Figure 2.10: Deep Recovery Mode helps students delve deeper into missing concepts.

they are struggling to master (Figure 2.9). Teachers see a notification on the Live Class Monitor, and the event is recorded in the student's personal data for the course.

- A student who continues to struggle transitions into Deep Recovery Mode (Figure 2.10). Here the content delves deeper into the concepts the student is missing. Now the vector has more force behind it – Vectored Instruction goes further back into the things the student needs to learn.

- A student who is still not progressing after Deep Recovery Mode transitions to Foundation Building Mode. This mode starts from scratch to build a whole new mental model of the concept that the student either learned incorrectly or missed.

When students break through and have the needed foundation, they emerge from Vectored Instruction precisely where they were in their course. Teachers see the transition on the Live Class Monitor. A student who has been progressing in the grade-level course suddenly comes to a standstill, is transitioned to Vectored Instruction, and after a time emerges right back where he or she left off in the curriculum empowered now to continue making forward progress.

Improving Memorization with Cognitive Thinking

Memorizing important information is critical to advancing to more difficult concepts. Examples include math facts and physics equations. Beginning readers need to memorize common words because the time needed to sound out words phonetically often degrades comprehension (Figure 2.11).

Figure 2.11: Learning is becoming easier and faster as scientists discover new methods for acquiring knowledge.

Brain imaging studies reveal that when information is repeatedly recalled, the brain eventually re-maps the memory by creating a shortcut. The desired state is long-term memory or instant recall. Traditional memorization techniques have their pros and cons for creating long-term memories.

Repetition

The traditional memorization technique is to set aside time every day for students to review the information that needs to be memorized, until eventually the brain re-maps the memory. This "brute force" method is extremely inefficient, doesn't work for some students, and is redundant and monotonous, leading to disengagement. In addition, when the brain is required to repeat the same task over and over again, it eventually falls into a lower state of activity.

Mnemonics

A newer method gaining momentum is creating mnemonics that pair an idea, such as a math fact, with a visual or auditory stimulus that is completely unrelated to the fact. These mnemonics, often humorous, are designed to create a lasting memory. (For example, "Kings play chess on fine green sand" for kingdom, phylum, class, order, family, species). The use of mnemonics has proven effective, especially for students who are unable to concentrate with the repetition method.

However, at a certain point the brain needs to re-map the memory – and that requires repetition. And sometimes, the lack of relationship between the mnemonic and the fact results in students being unable to recall the fact when the teacher isn't there to provide a hint.

The flaw with both methods – repetition and mnemonics – is that when the information is reviewed each day, it is stored in short-term memory. The brain draws upon short-term memory for the rest of the session, which means that the majority of the repetition cycles aren't exercising the part of the brain that creates long-term memories.

Spaced Repetition

Aristotle is credited with the concept that memory is created by the time lapse between direct experience and recollection.[13] This has spawned interesting studies on algorithms that use "spaced repetition" for memorization. Just as increasing the weight lifted increases muscle strength, increasing the time between reviews – say, from daily to twice weekly to weekly – strengthens memories. Students spend less time memorizing – and the time they do spend is more effective.

[13] *"...to remember, strictly and properly speaking, is an activity which will not be immanent until the original experience has undergone lapse of time."—Aristotle, On Memory and Reminiscence, ~500 B.C.E.*

Like other memorization techniques, spaced repetition also requires brain memory re-mapping – that is, creating a shortcut to the information in the brain.

Skirting Working Memory to Create Long-Term Memories from the Outset

Working memory is the part of short-term memory that processes information and forms thoughts (Sweller 1988, 1989, 1994). It is where most of our thinking occurs, and yet it has very limited capacity. In one of the most highly cited psychology papers, Harvard professor George A. Miller observed that humans can hold an average of seven plus or minus two chunks of information in working memory.[14] Presented with more than this many chunks, the brain's ability to absorb new knowledge begins to decline because working memory has reached its capacity. This capacity is referred to as cognitive load. The implication for learning? The more information that students need to store in working memory, the less new information they will understand and retain. Therefore, a student who has not already memorized some of the information referenced in a lesson will have difficulty following new information. The student has exhausted working memory capacity on information that other students have already moved into long-term memory.

What if there were a memorization method that could a) keep the brain from falling into a low activity mode (not a good thing in the middle of a learning session) and b) bypass working memory to create long-term memories that do not require conscious thought? This is called *implicit memory*, also known as unconscious or automatic memory. Examples are recalling the words of a song after hearing the first lines, or immediately recalling the product of 12 x 12 without having to work it out.

Implicit memories having the greatest impact require information to be meaningfully organized. When Dmitri Mendeleev organized the elements into the periodic table, something incredible happened. Not only was the information visually understandable for the first time, but Mendeleev was also able to identify holes: elements that had not been previously discovered. Innovative organization changed chemistry as we know it today.

Could a similar approach be applied to basic information to facilitate memorization? Consider the most commonly used table of addition math facts, organized by one of the digits in the problem (Figure 2.12). Notice two things: First, all of the math facts

---

[14] *"George A. Miller, "The Magical Number Seven, Plus or Minus Two: Some Limits on Our Capacity for Processing Information," Psychological Review, 1956*

appear twice. Does it really help a student to memorize 6 + 4 = 10 as well as 4 + 6 = 10? Second, there is no visual connection between 5 + 5 and 6 + 4. That's a missed opportunity for learning because students form memories the best when they can connect them to existing memories.

Addition table of 1	Addition table of 2	Addition table of 3	Addition table of 4	Addition table of 5
1 + 0 = 1	2 + 0 = 2	3 + 0 = 3	4 + 0 = 4	5 + 0 = 5
1 + 1 = 2	2 + 1 = 3	3 + 1 = 4	4 + 1 = 5	5 + 1 = 6
1 + 2 = 3	2 + 2 = 4	3 + 2 = 5	4 + 2 = 6	5 + 2 = 7
1 + 3 = 4	2 + 3 = 5	3 + 3 = 6	4 + 3 = 7	5 + 3 = 8
1 + 4 = 5	2 + 4 = 6	3 + 4 = 7	4 + 4 = 8	5 + 4 = 9
1 + 5 = 6	2 + 5 = 7	3 + 5 = 8	4 + 5 = 9	5 + 5 = 10
1 + 6 = 7	2 + 6 = 8	3 + 6 = 9	4 + 6 = 10	5 + 6 = 11
1 + 7 = 8	2 + 7 = 9	3 + 7 = 10	4 + 7 = 11	5 + 7 = 12
1 + 8 = 9	2 + 8 = 10	3 + 8 = 11	4 + 8 = 12	5 + 8 = 13
1 + 9 = 10	2 + 9 = 11	3 + 9 = 12	4 + 9 = 13	5 + 9 = 14
1 + 10 = 11	2 + 10 = 12	3 + 10 = 13	4 + 10 = 14	5 + 10 = 15
Addition table of 6	**Addition table of 7**	**Addition table of 8**	**Addition table of 9**	**Addition table of 10**
6 + 0 = 6	7 + 0 = 7	8 + 0 = 8	9 + 0 = 9	10 + 0 = 10
6 + 1 = 7	7 + 1 = 8	8 + 1 = 9	9 + 1 = 10	10 + 1 = 11
6 + 2 = 8	7 + 2 = 9	8 + 2 = 10	9 + 2 = 11	10 + 2 = 12
6 + 3 = 9	7 + 3 = 10	8 + 3 = 11	9 + 3 = 12	10 + 3 = 13
6 + 4 = 10	7 + 4 = 11	8 + 4 = 12	9 + 4 = 13	10 + 4 = 14
6 + 5 = 11	7 + 5 = 12	8 + 5 = 13	9 + 5 = 14	10 + 5 = 15
6 + 6 = 12	7 + 6 = 13	8 + 6 = 14	9 + 6 = 15	10 + 6 = 16
6 + 7 = 13	7 + 7 = 14	8 + 7 = 15	9 + 7 = 16	10 + 7 = 17
6 + 8 = 14	7 + 8 = 15	8 + 8 = 16	9 + 8 = 17	10 + 8 = 18
6 + 9 = 15	7 + 9 = 16	8 + 9 = 17	9 + 9 = 18	10 + 9 = 19
6 + 10 = 16	7 + 10 = 17	8 + 10 = 18	9 + 10 = 19	10 + 10 = 20

Figure 2.12: Addition Table

Now consider what happens when the facts are grouped by the sum (Figure 2.13). This organization facilitates memorization in two ways: reducing the number of math facts the student needs to learn and grouping facts in subsets that allows subconscious connections to be made – for example, the connection between 4 + 6 and 5 + 5. The student can draw upon better-known facts to help recall the less-known facts. This method has been proven effective but is traditionally used only to teach the sums of ten.

									5 + 5										
							4 + 4	4 + 5	4 + 6	5 + 6	6 + 6								
					3 + 3	3 + 4	3 + 5	3 + 6	3 + 7	4 + 7	5 + 7	6 + 7	7 + 7						
			2 + 2	2 + 3	2 + 4	2 + 5	2 + 6	2 + 7	2 + 8	3 + 8	4 + 8	5 + 8	6 + 8	7 + 8	8 + 8				
	1 + 1	1 + 2	1 + 3	1 + 4	1 + 5	1 + 6	1 + 7	1 + 8	1 + 9	2 + 9	3 + 9	4 + 9	5 + 9	6 + 9	7 + 9	8 + 9	9 + 9		
0 + 1	0 + 2	0 + 3	0 + 4	0 + 5	0 + 6	0 + 7	0 + 8	0 + 9	0 + 10	1 + 10	2 + 10	3 + 10	4 + 10	5 + 10	6 + 10	7 + 10	8 + 10	9 + 10	10 + 10
1	2	3	4	5	6	7	8	9	10	11	12	13	14	15	16	17	18	19	20

Figure 2.13: Pyramid of Sums

How Acellus Puts Cognitive Science to Work

Acellus presents lessons in a way that fills up the student's working memory so that the new bits of knowledge go directly into implicit memory. Although the focus of the activity is not the new knowledge, the student needs to repeatedly access the underlying information. This means the information must be organized in a way that doesn't require cognitive thinking to process.

Here's an example. Instead of simply entering answers to math facts, students are assigned to pair numbers that add up to a specified sum (Figure 2.14). After realizing that the sum is the same for the whole session, students can transfer their focus from the math facts to the cognitive task of pairing numbers. The pairing exercise occupies working memory requiring the student to subconsciously recall the math facts from their implicit memory.

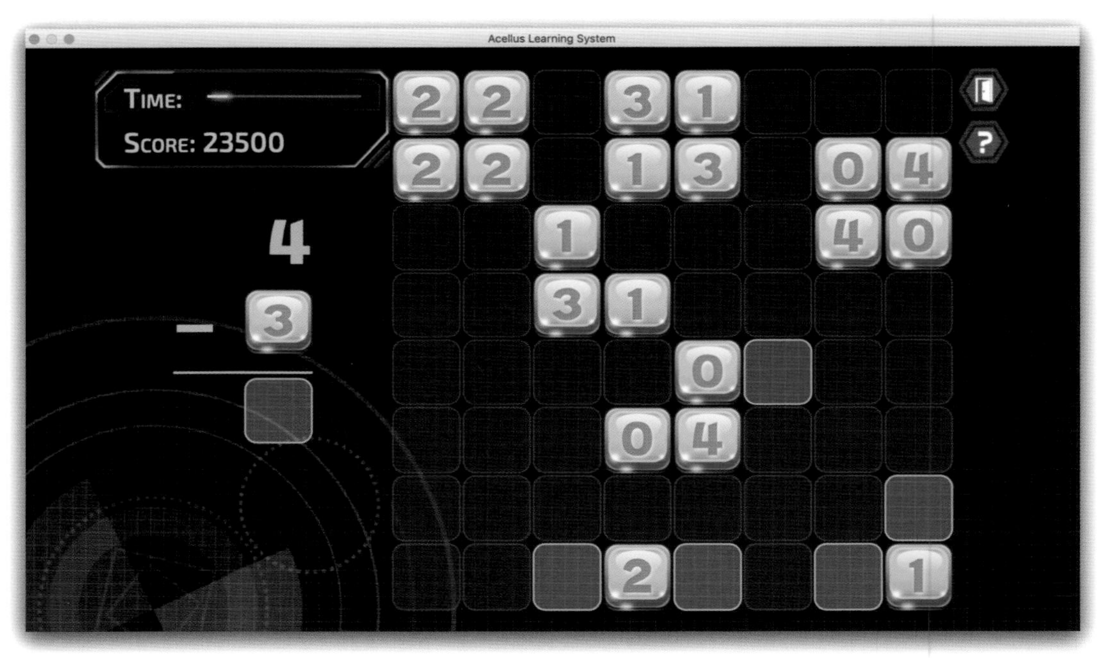

Figure 2.14: Acellus Math Facts Drill

In summary, Acellus continually explores new technologies, applications, research, and techniques to cater to the unique learning needs of every student. Standing on the shoulders of giants in education and innovation, Acellus gathers in the best resources, examines the strengths and shortfalls of new approaches to education, and then goes outside the box to discover and build solutions that deliver real results.

Chapter 3:
DEVELOPING ACELLUS COURSEWARE

Use the Acellus Courseware Development System to create the course syllabus, problems, textbooks, and lesson videos. This chapter provides guidelines and best practices for creating an Acellus course, from filming engaging videos to writing meaningful assessments and textbooks.

Developing Acellus Video Lessons

Acellus course content is focused on standards-based concepts. A *concept* is the point the teacher wants to convey to the student. A *lesson* is everything Acellus does to teach, reinforce, and help the student learn and master a concept.

Guidelines

Acellus course developers have a great deal of leeway but should follow these guidelines to maximize course effectiveness.

Do:

Teach one concept per video, or a few concepts that are closely connected. Include related ideas in the video if they reinforce the main lesson concept. Deliver the instruction students need to master the course material, freely combining lecture, multimedia, animations, and audio (Figure 3.1). Lessons give the on-camera teacher an opportunity to engage students in a concentrated way, creating the feel of a one-on-one learning experience.

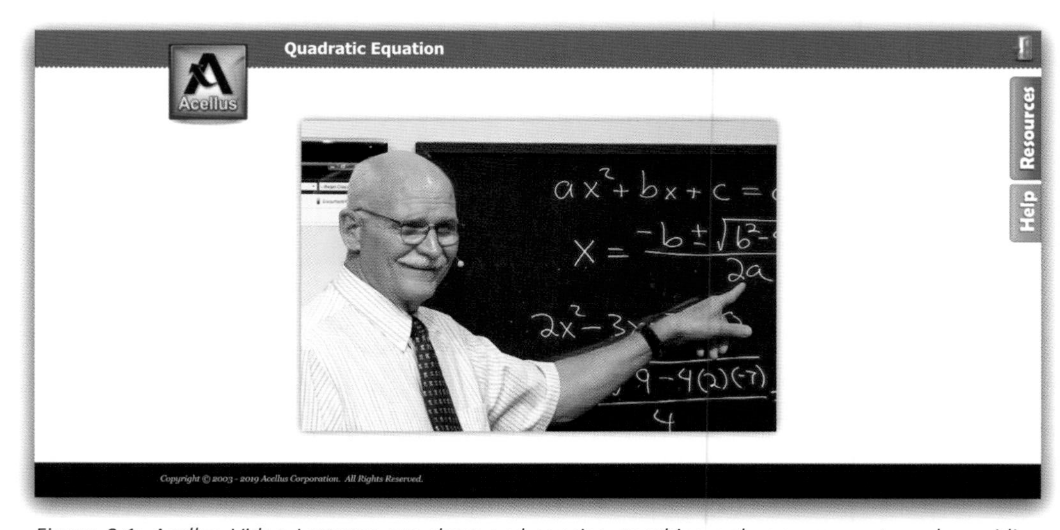

Figure 3.1: Acellus Video Lectures are short and concise, teaching only one concept, and providing students with only the bits of knowledge that pertain to that concept.

When appropriate, create multiple videos to teach the same concept at a different level or from a different angle:

- Level A videos are targeted to average learners.
- Level B videos are targeted to students who need a little longer to master a concept. These videos often include extra examples. When Acellus recommends a student be moved to Slow Mode for the course, this is often the video shown.
- Level C videos are targeted to students who continue to struggle with a concept even after being shown the Level B video. These videos review concepts taught in previously lessons to lay the foundation for the current lesson. Sometimes they cover the concept in a more basic manner.
- Videos for Accelerated Mode are targeted for gifted students. They provide a more rigorous treatment of the concept.

Make the video as long as necessary to teach the main concept – and no longer. Following are guidelines:

- Elementary-level courses: 3 to 7 minutes
- Middle school-level courses: 5 to 10 minutes
- High school-level courses: 7 to 15 minutes
- AP-level courses: 15 to 25 minutes

Present lectures in the style that's most effective and natural for you. On-camera teachers have latitude in terms of concepts to cover, the order of concepts, and books and materials used as resources for preparing the lesson content.

Follow any required curriculum standards, as dictated by the Acellus Editorial Board. While the on-camera teacher is considered to be the expert on teaching the material, the Acellus team provides constructive feedback to help make the teaching experience as professional and effective as possible.

Use visuals and demonstrations to enrich the learning experience. Consider including PowerPoint slides, white boards, or animated text or images added by our editing team after filming. A word of caution: when the lesson progresses very slowly – for example, when a teacher is writing on the board – students tend to get distracted and lose concentration. Consider adding animated writing (which is faster) to the video in post-production. Feel free to request graphics or animations that you feel would enhance a given lesson. Either submit requests to the editing team in writing or state the request in the video before the start of a lesson – for example, "Note to editors: In this lesson I will refer to cities along the Nile River – please display a map of the river and then highlight the cities as I refer to them." Animated text or images added during post-production can be presented as floating graphics that share the screen with the teacher or else fill the entire screen.

Look for ways to engage students. Effective techniques we've seen:

- Introduce a surprise. A surprise that reinforces the concept being taught is even better.
- Include examples of student interests that relate to the concept.
- Start lessons in unexpected ways to grab the student's attention and interest. Consider a fun prop, short story, interesting question, or unexpected entrance.
- Speak in a conversational and personable manner. Avoid speaking slowly and formally: studies show this can decrease student engagement.
- Look at the camera. Without actual students in the room it can be tempting to teach to the floor or the wall. When you look at the cameras, students feel you're looking directly at them.
- Include students in the video if they will enhance the learning experience. Most Acellus videos are filmed without students in the studio.

Be mindful of the needs of diverse learners when preparing and filming course content. Acellus courses are offered at:

- Public schools
- Charter schools
- Private schools
- Faith-based private schools
- Distance learning schools
- Home schools
- Libraries
- Detention centers

Do Not:

Date content. Be careful to not tie the course to a specific time period by discussing current news events or by using phrases such as yesterday, tomorrow, last year, next year, and so on. When years are needed – for example, when teaching students how to write a letter – we suggest using dates at least five years in the future.

Tie lesson to a location. The student should feel like the on-camera teacher is where the student is.

Tie lesson to a particular time of day or season. Don't begin a lesson with "Good morning" or "Good afternoon" or mention the weather outside today unless it is relevant to the lesson.

Tie lesson to a particular age group. Rather than saying, "This is Freshman Biology," say, "This is Biology." Older students may be taking the course.

Offend any part of the target audience. Cover sensitive subjects with respect, taking care to not insult any viewers' moral beliefs. When discussing controversial topics, cover varying viewpoints in an unbiased manner.

Ask students to do something they may not be able to do. Not every student can go to the library or browse the Internet because they may not have access. Not every student can get permission from their parents because they may not have parents or their parents might not be local. Better to say, "You will need adult supervision."

Assume that materials and supplies are readily available to students. Instead, bring your own supplies to the studio and model the activity for the students.

Refer to previous or future content. The order of lessons in a course may later be shuffled. Avoid statements like, "This is the last lesson in this Acellus course."

Refer to your lecture as a "video." Students should feel like they are being taught one-on-one and that the teacher is right there connecting with them. Phrases like "in this video" call attention to the fact that the lesson is pre-recorded.

Developing an Acellus Course

Step 1 - Generate Course Curriculum (Lesson List) for Approval

Standards-based textbooks serve as excellent references for preparing the curriculum. Start with a list of lessons. Limit each lesson to one concept (or closely related concepts) even if the material can be covered in just a few minutes. One lesson per concept. Each concept is something that the student does not know yet and should understand after watching the video.

Give each lesson a unique title that clearly reflects the concept. Write a concept description – a concise sentence describing the concept to be taught. This sentence should convey the focus of the lesson. The on-camera teacher refers to the concept description when filming to keep the video lecture on track. Document the specific state standards that the lesson satisfies.

Submit the lesson list, associated concept descriptions, and satisfied standards to the Acellus Curriculum Board for approval.

Step 2 - Prepare Lesson Content, Following the Acellus Teaching Model

To create effective, engaging, and empowering lessons, follow the Acellus teaching model shown in Figure 3.2.

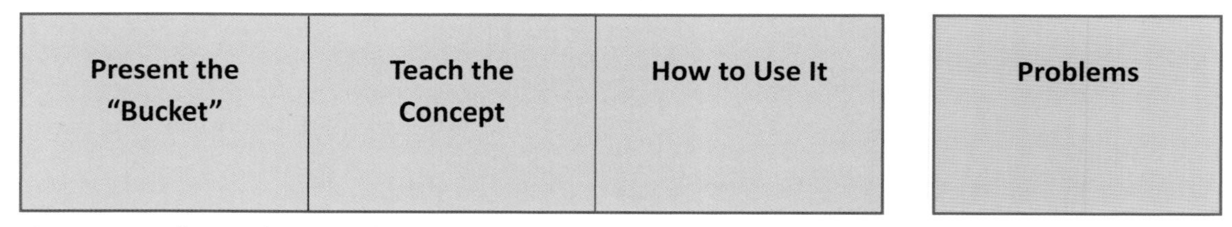

Figure 3.2: Acellus Teaching Model

1. **Present the bucket.**

   Provide context for the lesson: the concept along with a higher-level view. This helps students understand how the new content hooks on to what they already know, making it relevant.

2. **Teach the concept.**

   Present the nuts and bolts of the concept. Teach all concepts normally covered in the classroom, focusing on the content required by standards and omitting content that is irrelevant or could be confusing. After explaining the concept, provide relevant examples that prepare students to solve the problems they will be assigned after the video.

3. **Explain how to use the knowledge.**

   Help students see how they will be able to use the new knowledge. Some concepts simply provide the foundation to learn other concepts, and students will need to learn several concepts before they can do something with the information. In these cases, explain to students how they will eventually be able to use this information. Example: "This is one of the things that you will need to know to be able to build a rocket."

   In other cases, students will master the content sufficiently to apply this understanding to solving everyday problems. In such cases, consider creating a separate video for the applications of the concept. Breaking this material out into two or three videos allows students to focus on different aspects of working with the concept after each video lesson.

4. **Create practice and assessment problems.**

   Ideally, problems should follow every video. Someone other than the on-camera teacher often writes problems. Tips for writing effective problems appear later in this chapter.

Step 3 - Film the Video Lesson

When filming a lesson, think about the target audience to make each concept engaging, interesting, and exciting. Students should feel as though the on-camera teacher is right there teaching them. Be sure to keep in mind the focus of each lesson (as stated in the concept description) and present each concept as concisely as possible to keep the student's interest.

Todd Edmond
On-Camera Teacher

Teaching On-Camera!

Lights...

Camera...

TEACHING?

There are some teachers who claim they are not entertainers. However, there is no doubt that a profession in education requires the ability to perform in front of others – in this case, students. The educator that is in front of the learner is responsible for the way the subject matter is presented, and for holding the attention of the students themselves.

The Acellus filming system is very unique. It requires a teacher to step out of their comfort zone and away from a functioning classroom. The task can be overwhelming at first – often with the teacher having no idea how to instruct without actual students in front of them! However, with proper planning and preparation, teaching seven-minute lessons to a camera can be both enjoyable and eye opening as an educator.

Once the outline is complete, an Acellus teacher can get to work on the "meat and potatoes" of the course – which also tends to be the greatest amount of work. The instructor needs to determine what needs to be taught, how best to teach it to a camera, and what visuals (if any) need to be presented. Research, time, and the preparation of materials is all a part of this phase. The teacher is the expert!

The goal is to have everything understood and ready to go by the time the teacher hits the studio for filming. The instructor should be a master of the material before the camera turns on.

This will allow for better flow to the lesson, and for the educator to *teach to the student*, instead of trying to remember what to say.

Filming in the Studio

When the time comes to film the course, an Acellus teacher should be completely ready to go. All presentation materials should be in the system. It is always a good idea to have either a hard copy or your laptop ready in the event you need to freshen up on each lesson before the camera turns on.

The key to reaching students is threefold:

First, you must be relaxed and not robotic in front of the camera. Students will become bored very quickly if you are simply stating facts or are going through the motions trying not to mess up. Imagine you are in your classroom with students in front of you. Be lively! Smile! Act as though the kids are in the studio with you. The camera has become the face of your students! Enjoy the moment of teaching. Enjoy the subject you are presenting in the lesson. Show students your passion and love for teaching.

Second, it is okay if you make a mistake! This is the world of filming. If something goes wrong you simply need to start over and do it again. It will seem very weird at first to speak into a camera in order to teach. However, once you realize that you are still educating – that there are kids out there watching you – the flow of a lesson will come through. It may take a few lessons to get the feeling of talking into a camera down, but mistakes are part of the process. One take is not required! Relax and teach.

Finally, any good Acellus teacher must be personable in order to get through to the kids they are teaching. The material and curriculum are important. The students need to understand and grasp the concepts in each lesson. However, without a human being actually teaching it to them, it becomes lifeless. Add humor to your lessons. Make eye contact with the camera on a regular basis. Come up with a catch phrase to end each lesson such as, "see you next time!" The more personable you make your lessons, the more students you will reach.

Passion Makes Perfect

Every great educator loves to teach, and it shows. Students who walk into their classrooms (or their "studios") will automatically notice that they have a passion for their subject, for their teaching, for their students.

Think back to those teachers you yourself had growing up. Ask yourself why those teachers made learning so much fun. What did they do to make you want to come back again and again, day after day? Write those reasons down, and apply it to your teaching at Acellus.

The kids are still there with you. They are watching, listening, and learning everything you present. It may seem lonely and different in a studio, but the students are everywhere. Look into the lens, think of the kids, and do the job you have always done. In the end, great courses will be created for thousands of learners waiting for your instruction and passion!

Empathy

Sense of Community

Positive Reinforcement

Innovative Pedagogical Techniques

Mark Rogers
On-Camera Teacher

Empathy

Excellent teaching is an exercise in empathy – can you dig into your memories, close your eyes, and remember your feelings and perceptions in different classrooms throughout your academic career? Take a moment to remember your favorite teachers and classroom moments.

Did you remember any of the following?

1. An educator you felt truly cared for your well-being, future, and growing skill set
2. An environment that felt safe but had a whimsical fun that made attendance enjoyable
3. A predictable routine that paradoxically also contained moments that were completely unexpected

Why do we start with empathy? Putting ourselves into a situation where we feel the way someone else does allows us to truly understand what it's like to crawl inside that person's brain.

So, what does an Acellus student feel? What type of situation might they be in?

1. *Homeschool* – they may be taking courses through Acellus in lieu of attending a brick-and-mortar school. This can help with advancing through content faster, but we must be aware of their desire for connectedness and community – there are no other live students with which to interact.
2. *Credit Recovery* – they may be taking courses through Acellus

because they did not earn a passing mark from their school's classroom teacher. We must be aware that they may feel ashamed of this process – they are in need of positive reinforcement along with unambiguous, clear instruction so that their confidence may be rebuilt.
3. *Blended Learning* – they may be taking courses through Acellus because the school employs a blended learning approach – some content will be taught live through in-person teaching and Acellus may be the enrichment or supplement to this live instruction. The kids need another angle with which to learn the content – teaching them using unique examples and innovative styles will pique their interest.

Through this empathy exercise, we have identified the three main pillars with which Acellus teachers must approach in-studio recorded lessons:

1. Building a sense of community
2. Positive reinforcement
3. Innovative pedagogical tactics

Building a Sense of Community

The course is seen as a community effort where success is a team effort. Pronouns "we" and "us" are essential so that it actually feels like a multi-student classroom.

Positive Reinforcement

The student is frequently commended for work completed throughout the lesson, even though the in-studio teacher cannot interact or see the student in that moment. "Excellent work," and "that was challenging, but I knew you could do it" are just two ways you can develop rapport and build the confidence of students taking Acellus courses.

Innovative Pedagogical Tactics

Teaching the material in ways that are new, exciting, and innovative are the surest way to engage the widest group of students, no matter the situation in which they find themselves watching a course video. Fortunately, in-studio teachers have a talented team employing cutting-edge technology to support them in this quest. On-site filming in locations around Kansas City, a green screen to put virtually any background behind the on-camera teacher, and a set replete with props mean that the only limiting factor is your imagination. Think long and hard about how each lesson could begin with engagement, and you will assuredly have buy-in from your student to learn the content with an engaged mind.

Example:
On-camera teacher wants to film a video about taking the percentage of numbers.

Standard approach: "Welcome students. Today we're going to learn about taking the percentage of numbers. On the board I have 10. Let's say I want to take 40% of 10. I need to convert 40% into a fraction or a decimal first; 40% is 40/100. Now I can multiply 10 by 40 to get 400. Now I divide by 100. My answer is 4. 40% of 10 is 4. Thanks for watching."

The Acellus approach: "Welcome to my local grocery store, students <yawns>. Now one thing you need to know about me, students, is that I have to have my coffee in the morning <yawns again>. And I'm out of those magical roasted beans! So here I finally am, walking down the coffee aisle, trying to find the best deal. $12, $14, $25! Wow, coffee can get EX-PEN-SIVE! Ooo, here we have another one... $10. AND it has a 40% discount! Now I know the word "discount" means that we won't need to pay full price, but what does that mean exactly? How would we calculate that? Can you help me?"

<returns back to studio for lesson>

"Oh hi, students, welcome back from the grocery store. I appreciate your help in finding a cheap way to get me caffeinated. We were looking at those $10 coffee beans with a 40% discount. When we want to take discounts, the first step is to convert the percentage into a fraction or decimal. That's right, you can really

choose whichever one gives you the greatest inner sense of peace <slowly raises hands up and down and breathes in and out slowly>. Yes, I do believe that the fraction gives me the greatest inner sense of peace today. I know 40 per-cent literally means 40 parts of one 100 because the root word “cent” means 100. So, we will take 40/100 and multiply that by our target number, in this case $10. Multiplying fractions means we multiply numerator by numerator, so 40 times 10 will give us 400.

Now our denominator is still 100, so our last step to calculate this discount will be to divide 400 by 100. There are 4 groups of 100 in 400, so our answer is 4. But just wait, applied to our coffee problem, that means our discount is 4, so if our original price is $10, and our discount is $4, that means that the coffee will only set me back $6.

Oh, the sweet, sweet smell of a cup of coffee. Thank you for helping me today; you did great!”

After reading the standard approach and Acellus approach above, please think how the on-camera teacher emphasized each of the following pillars of Acellus teaching:

1. Building a sense of community
2. Positive reinforcement
3. Innovative pedagogical tactics

You, the on-camera teacher, were chosen to join the team because you have impeccable content knowledge and a knack for reaching kids. Use your skill set and creativity to build lessons that will empathize with the students. You don’t know what happened before they pressed play, and you don’t know what will happen after they log out, but you can control to a large degree how they feel about your course. Make the difference for them; they’ll love you for it.

Developing Acellus Help Videos

Short Help Videos help students learn how to solve a specific type of problem. Students see a sample problem being worked out as they hear an explanation (Figure 3.3). These videos are filmed using a document camera or by the on-camera teacher in an Acellus studio. Help Videos are particularly helpful to students who are stuck on the problems and need a quick, focused review on how to solve them.

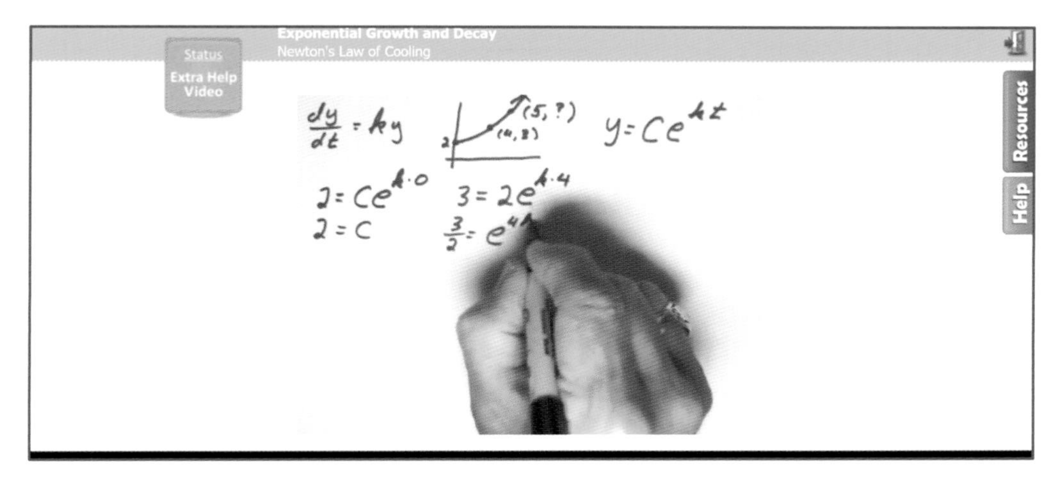

Figure 3.3: Help Videos provide students with an example of the kind of problem they are learning to solve – and how to solve it.

Writing Acellus Problems

After completing a lesson video, students are given problems to assess their understanding of the material and to add rigor to the course. The problems are a crucial component of Acellus courses, being used for student evaluation and to provide feedback for Acellus developers so that they can continually improve lesson effectiveness.

The multi-mode capability of Acellus allows teachers to tune lessons to each student's individual needs and abilities. Students in Accelerated Mode work more problems than students in Normal Mode, and the problems may also be more challenging and sophisticated. Students in Slow Mode work fewer problems, which may also be simpler. These students often need more time to solve problems, and reducing the number of problems enables them to complete course requirements in the allotted time.

Problem Types

Create problems in the Acellus Courseware Development System, designating each one as an instructive, practice, or assessment problem. A single problem can be designated as belonging to one, two, or all three categories.

Instructive Problems

Instructive problems are presented to every student in the order specified in the Acellus System (Figure 3.4). Use them in the following ways.

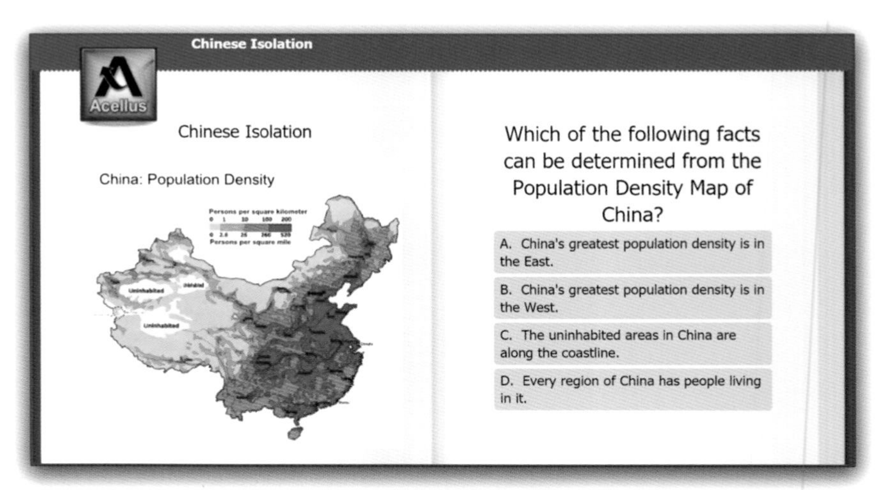

Figure 3.4: Instructive Problems can be used to reinforce concepts that have been taught, to lead students incrementally through complex concepts, to review lesson concepts, and to stimulate cognitive thinking.

Reinforce concepts taught. Students often grasp concepts during the teacher's lecture only to find them slip away when it's time to start working the problems. Instructive problems help reinforce lesson concepts. One way to present instructive problems is to summarize lesson content in the left column and present related problems in the right column. Referring to the content while solving the problem helps students review and master the lesson concept.

Incremental Complexity. Consider creating instructive problems that increase in difficulty. If a problem involves multiple steps, for example, you might create one or a set of problems that deal with just the first step. The next instructive problem (or set of problems) can focus on the first and second steps. The last group of problems can integrate all steps. After mastering instructive problems, students are ready to tackle the lesson problems for the concept.

Review lesson concepts. Instructive problems can also help students review the material covered in the lesson. Craft these problems to help students rediscover concepts learned during the lecture.

Stimulate cognitive thinking. In school, work, and daily life, our senses constantly bombard us with information. It's up to the brain to receive, process, store, and access that information when we need it. The power to do this comes from the brain's core cognitive skills, including long- and short-term memory, auditory processing, visual processing, logic, and reasoning. All of these cognitive skills play an important role in processing new information. Even one weak skill impedes our ability to grasp, retain, or use incoming information.

The right instructive problems help to improve and develop a student's cognitive thinking ability. The approach involves combining audio, visual, or both types of stimuli with information the student has already learned, and then challenging the student to process all of this information into logical decisions or conclusions. Story problems are often used in math to stimulate the development of cognitive thinking.

Acellus course developers can expand this process. An effective technique is to combine information learned in previous lessons (stored in long-term memory) with material learned during the most recent lesson (stored in short-term memory), and then present an audio or visual stimulus. The student is challenged to arrive at a conclusion that is only possible by considering all three inputs. Learning to combine information from multiple sources to reach a non-obvious conclusion is an effective way to practice and develop cognitive thinking skills.

Prism Diagnostics Problems

Prism Diagnostics problems are a special type of instructive problem designed to discover common misconceptions. To study the learning process and improve course effectiveness, researchers at the International Academy of Science analyze student response data stored on the Acellus Courseware Development System. It's not uncommon that a notable percentage of students enter the same identical wrong answer to a specific problem. By examining the problem more closely we can uncover the common misconception. In these cases we film a deficiency recovery video that begins as soon as a student enters a common wrong answer, helping to correct minor misconceptions.

It is possible to write special problems that reveal deficiencies in the student's background regarding a lesson concept. To write "prism" problems, researchers begin by analyzing the lesson concept to predict common student misconceptions. Let's say, for example, that the concept is to add and subtract fractions. Some students may not know that a common denominator is needed. Others don't know how to find a common denominator. Still others may not understand equivalent fractions. Consider writing one or more prism problems for each deficit. Students who answer in one of the wrong ways predicted are shown the deficiency recovery video, giving them the specialized personal instruction they need to succeed.

Practice/Assessment Problems

Practice problems test students' understanding of the lesson concepts and add practice or rigor to the course. The course developer can designate some or all practice problems as assessment problems for use in exams.

Creating Effective Problems

Follow these steps to create effective Acellus problems:

1. **Identify the specific concept that the lesson intends to convey to the student.** In some cases, the on-camera teacher might incorporate several concepts into a video lecture to better illustrate, reinforce, and teach the main lesson concept. Even in these cases, make sure that the problems assess student understanding of *the single concept that is the focus of the lesson.* When it is important for students to practice solving problems that require skills acquired from previous lessons, consider creating separate lessons with their own video and problems.

2. **Select the best Acellus problem template to assess student comprehension of the presented concept.** Students respond by clicking the correct answer or entering it on the keyboard. Templates are available for:

   - Video problems – The on-camera teacher asks the student a question requiring a response.
   - Dynamic graphics – Students respond to a question about a graphic, with or without an associated audio clip.
   - Static graphics – Students respond to a still image, with or without associated audio clips.
   - Reading assignment – Students see a scrollable narrative along with associated questions.
   - Percent accuracy problems – These problems accept answers within a specified percentage of the acceptable answer. This type of problem is especially useful for problems where rounding differences may result in a slight variation in student responses.

3. **Select the best problem format.** We recommend multiple choice problems, simple answers, or step problems. Step problems provide step-by-step prompts to the student, enabling them to provide answers that are difficult to enter from the computer keyboard.

   Avoid true/false problems. They are allowed but are inadequate for assessing students' depth of knowledge.

4. **Write problems, following these guidelines:**

   - Write each problem based on the information provided in the video.
   - Make problems as short and concise as possible.
   - Avoid problems with more than a few correct responses.
   - Take care that all practice and assessment problems for a given lesson have the same complexity and difficulty. The reason: each student will be given a unique, randomized set of problems and their scores will be compared to students given a different randomized set. If you want to include problems with multiple levels of difficulty, create separate lessons.
   - There is no limit on the number of problems you can provide. Each student will be shown a random subset of the problems.
   - Take care that each assessment problem provides enough information for students to understand the question if it is not presented directly after the lesson. Assessment problems are used in unit, mid-term, and final exams.

5. **After writing the problems for a lesson, read through each one carefully, looking for any ambiguity.** Enter all the possible correct answers for each problem.

Writing Answers

Acellus compares student responses to the answers stored for that problem. Problems may have one or more correct forms of the same answer that are acceptable — for example, "one in ten" or "10%." Be sure to include all acceptable answers. Answers are not case-sensitive or space-sensitive.

Creating Acellus Special Lessons for Blended Learning Environments

Research shows that students taught in a blended learning environment – having access to a combination of online and face-to-face instruction – excel in relation to peers who have exposure to only one method of instruction. When implemented effectively, a blended learning program can make better use of instructional resources and facilities and increase content and course availability, speeding up students' pathways to graduation. Acellus provides the face-to-face instruction component of blended learning through Special Lessons (Figure 3.5).

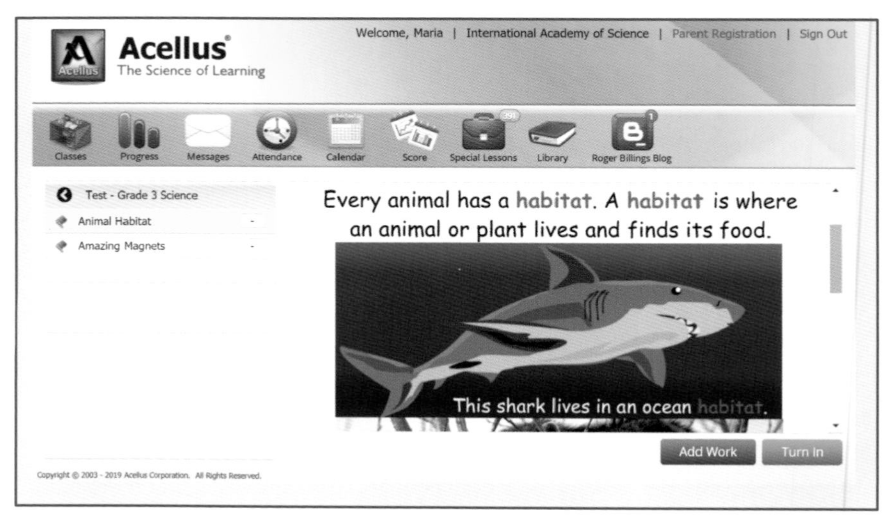

Figure 3.5: Acellus Special Lessons are provided as resources for the blended learning environment.

Classroom teachers use Special Lessons at their discretion. The lessons provide course-specific, teacher-driven classroom activities to enrich the learning experience. Activities range from writing assignments and in-class discussions to games and classroom experiments. Special Lessons also give students opportunities for teamwork and collaboration.

Acellus Special Lessons have two parts:

- The Teacher Resource provides the information and instructions needs to carry out the activity. It also contains a grading rubric, where appropriate, to assist the teacher in grading the assignment.
- The Student Resource presents the background information and details the student needs to complete the activity.

Write Special Lessons at the appropriate level for the target audience. High-resolution graphics and images are encouraged, especially in the Student Resource.

Writing Acellus Textbooks

Acellus textbooks provide a detailed and rigorous presentation of the subject material corresponding to each video lesson in a course. All Acellus textbooks are reviewed and critiqued by the Course Developer Community and the Editorial Board for quality, rigor, professionalism, and compliance with the textbook vision and guidelines.

Textbook lessons cover the same concept as video lessons but should stand alone even without the video lesson. Textbooks provide:

- An alternative method for a student to learn the lesson material.
- Help and support for a student who has watched a video lesson but is struggling with the practice problems.
- A method to review a concept in preparation for taking exams, highlighting key lesson points and relevant materials.
- More insight into the lesson concept.

Although Acellus textbooks are provided for all Acellus courses, they are especially important and effective for AP and college-level courses, where acquiring foundational knowledge sometimes takes more work and time to assimilate.

While textbooks teach the lesson concept in-depth, they are not necessarily limited to the material covered in the video. Nor do they need to follow the same teaching method as the video lesson. When authoring a textbook, freely include:

- Sample problems and examples.
- Related information and analogies to help students better relate to and understand the main lesson concept.
- Sidebars, featuring real-life connections and profiles of individuals making important contributions to the field of study.
- High-resolution images and graphics.

Acellus Lecture Notes are a special type of Acellus textbook that provide the content delivered in the video lecture. For some courses, Lecture Notes are the most effective way to present the content students are expected to master. When preparing Lecture

Notes, thoroughly cover the concept discussed in the video lesson, including any relevant examples, problems, and solutions.

Acellus Textbook Format

When writing an Acellus textbook, use these elements:

Lesson title: Place this at the top of the page in a larger font than the body copy.

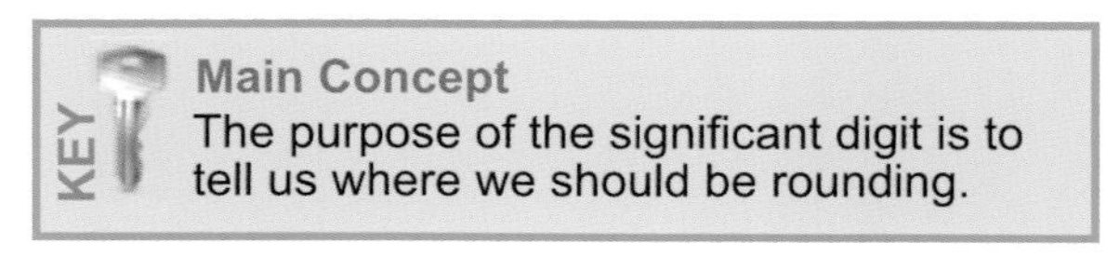

Figure 3.6: An example of a Main Concept Box

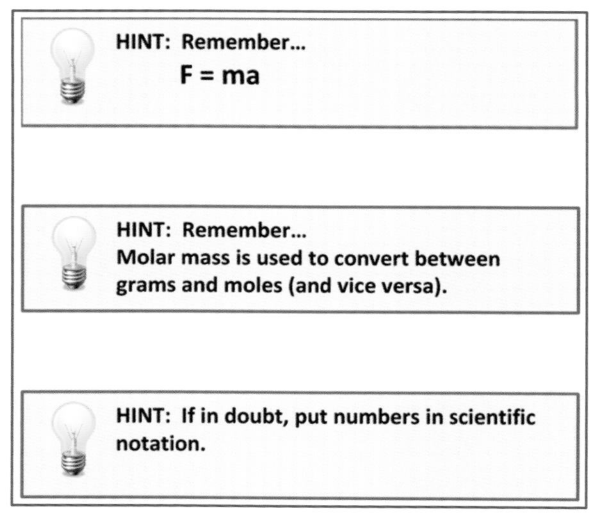

Figure 3.7: Three examples of Emphasis boxes

Main Concept: Optionally include the main concept of the lesson in a green box (Figure 3.6).

Emphasis boxes: Highlight important facts to remember by placing them in emphasis boxes, which include a light bulb and the word “Hint” (Figure 3.7).

Images: All images should be original or from the public domain. Make sure labels are legible.

Fonts: Color, font type, and size should be legible and age/grade appropriate. Use the same font type and size throughout the textbook, as much as possible.

Voice: We prefer that you write textbooks in active voice as opposed to passive voice. (Written in passive voice, the previous sentence would read, “It is preferred that textbooks be written in active voice.”)

Continual Improvement

New courses are published to the Acellus Library Server, where they are available to Acellus servers around the world.

As students work through the courses, the Acellus Courseware Development System collects and analyzes student response data to identify areas of difficulty. The remedy might be simply replacing an ambiguous problem. Or it might be filming a new video to teach the lesson concept in a different way. In the latter case, the Acellus on-camera teacher is invited back to the studio to film additional video content. The goal is continual improvement.

Chapter 4:

BECOMING FAMILIAR WITH THE STUDENT INTERFACE

To help their students succeed on Acellus, it is important for teachers to understand what Acellus looks like from the student's perspective. This chapter provides an overview of the Acellus Student Interface, accessible through the Acellus App (recommended, Figure 4.1), or through www.acellus.com (Figure 4.2) using any major browser. Students simply enter their Acellus ID and password to sign in.

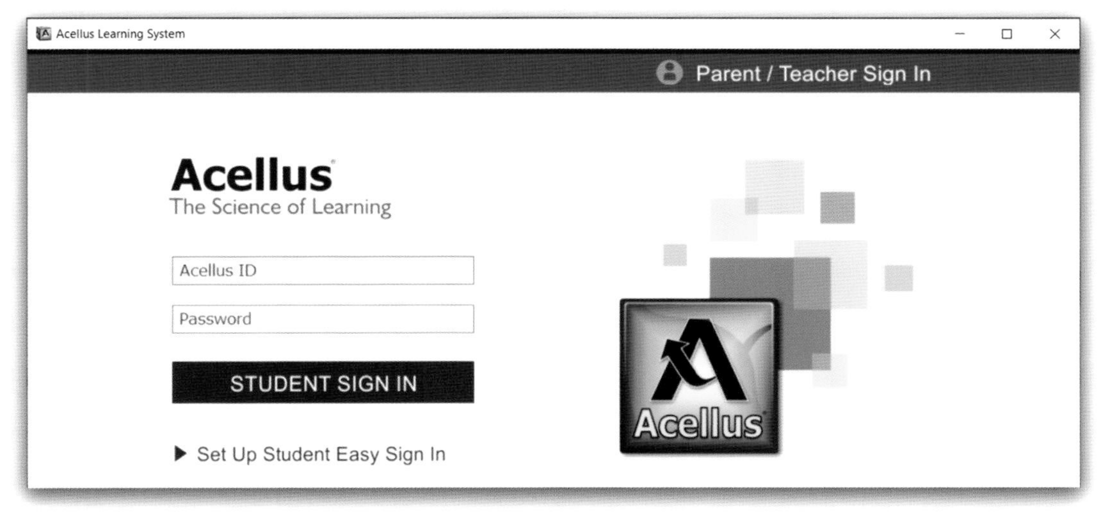

Figure 4.1: Students can sign into Acellus from any location that has Internet access by going to acellus.com. However, we recommend that students use the Acellus App, which can be downloaded from acellus.com – *select* Download the App *located on the bottom menu.*

Figure 4.2: Students can access the sign-in page by opening the Acellus App or by selecting Student Sign In *at* *www.acellus.com, as shown.*

The Student's Online Learning Experience

After signing in, students see their class list, shown in Figure 4.3. (Clicking on the Classes button from the menu bar also opens this page.) Selecting a class takes students to their current position in the online course.

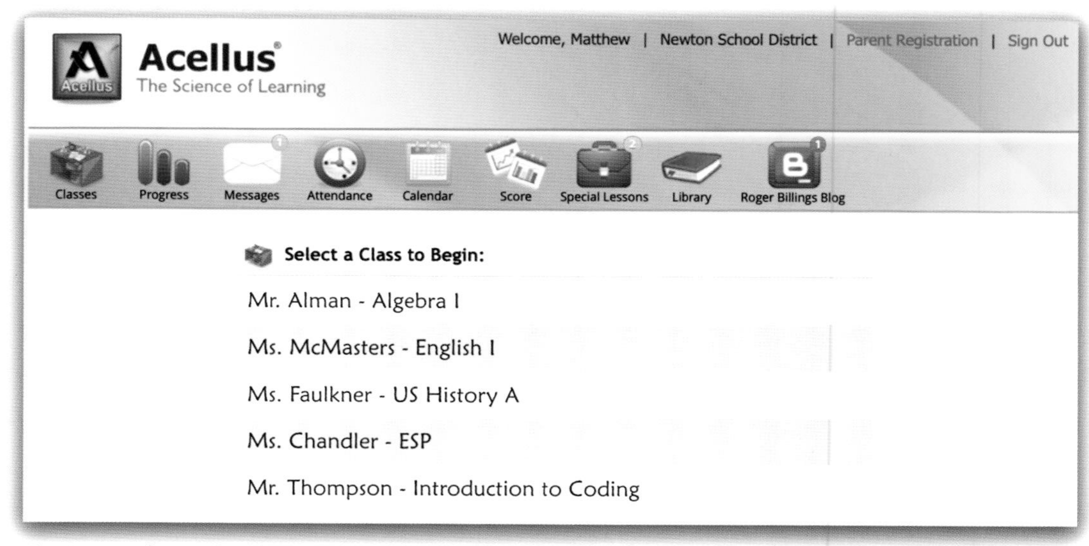

Figure 4.3: The student class list shows the classes that students are currently enrolled in. Students are also given multiple menu options to navigate the system.

Videos

Lessons begin with Acellus delivering video-based instruction. With every lesson focused on the instruction and mastery of a single concept, Acellus can help students build a solid knowledge base.

Save Video Position Feature

Sometimes students are in the middle of a video when they must stop their studies. With the Save Video Position feature, Acellus remembers their video position and allows students to sign out in the middle of a video and then resume where they left off the next time they log in.

Video Skip Back Feature

Students also have the option to use the skip back icon shown in the lower-left corner of the video screen (see Figure 4.4). This rewinds the video 15 seconds, letting them re-watch an important point they missed or want to reinforce. Students can click the button multiple times to rewind as far back as needed.

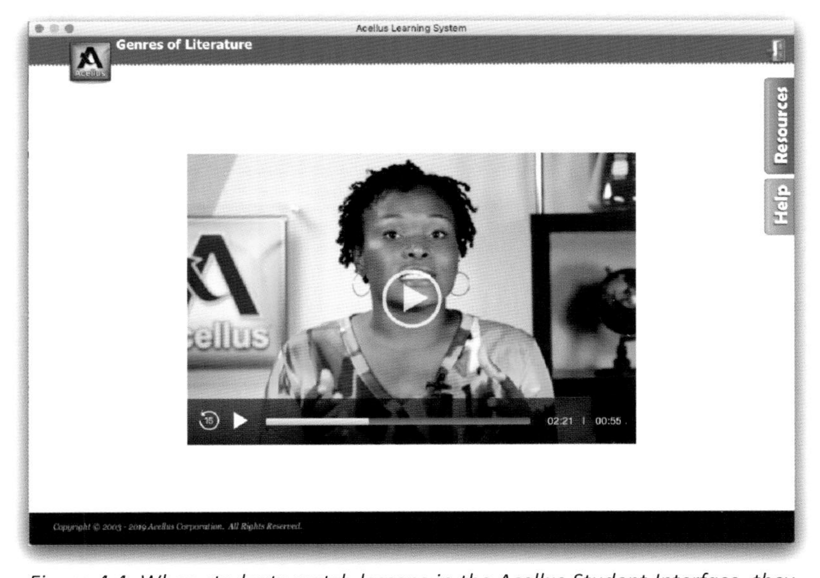

Figure 4.4: When students watch lessons in the Acellus Student Interface, they have the ability to skip back 15 seconds at a time to repeat important information. When students sign out in the middle of a video, Acellus remembers their position and starts the video where they left off when they return.

Problems

Practice Problems – to Assess Mastery

After the video, practice problems are given to assess students' understanding of the lesson concept (Figure 4.5). Using Prism Diagnostics and Vectored Instruction, Acellus diagnoses areas of student weakness and provides targeted help in these areas – customizing the learning process and filling in holes in students' understanding for a strong learning foundation.

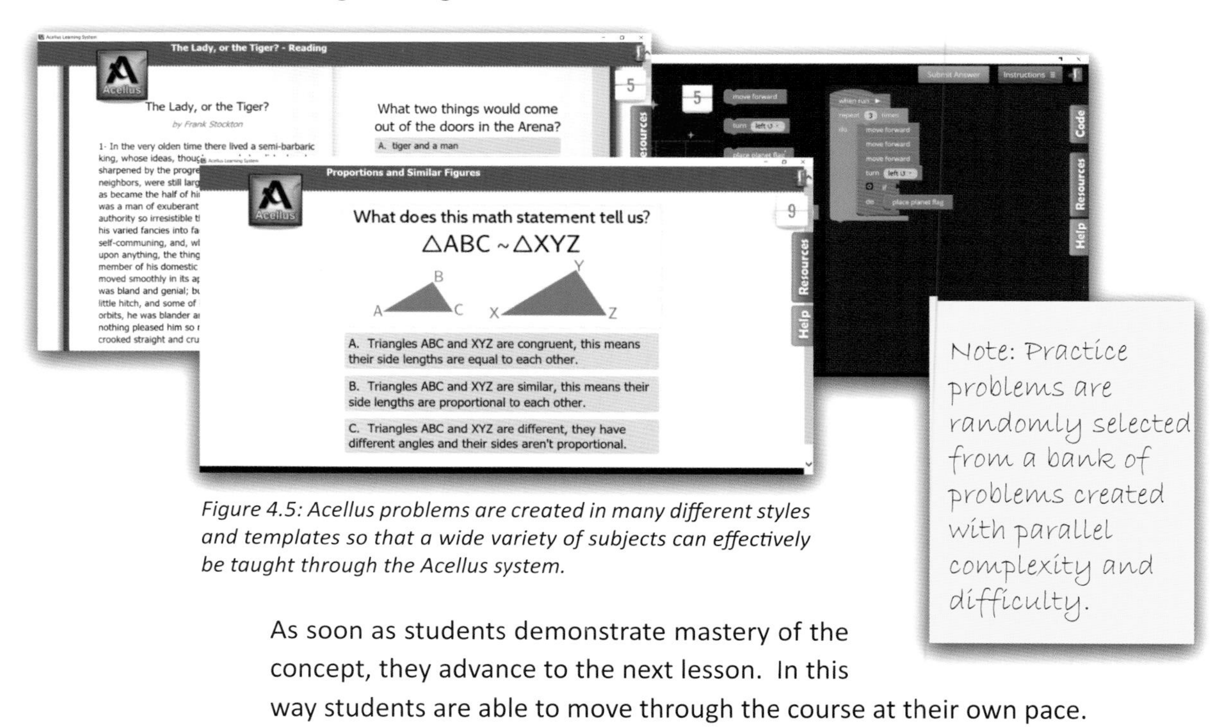

Figure 4.5: Acellus problems are created in many different styles and templates so that a wide variety of subjects can effectively be taught through the Acellus system.

Note: Practice problems are randomly selected from a bank of problems created with parallel complexity and difficulty.

As soon as students demonstrate mastery of the concept, they advance to the next lesson. In this way students are able to move through the course at their own pace.

Students can track their progress through lessons and exams with a problem counter which displays the number of practice or assessment problems remaining before they complete the step.

Instructive Problems

Note: Instructive problems are not included in the student's problem counter.

As students progress through their courses, they will sometimes encounter instructive problems. These special problems reinforce the lesson concept and prepare students for the practice problems that follow.

Unlike practice problems, which are randomly selected for each student, instructive problems are presented to every student in the order designated in the Courseware Development System.

Help Tab

The Help tab is a resource students can use if they need extra help while solving problems. Here they have access to any help resources available for the course. From the menu (shown in Figure 4.6) students can select from the following:

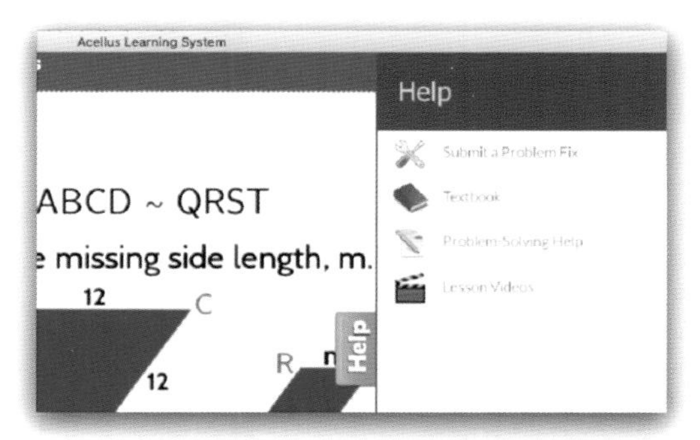

Figure 4.6: The Help tab gives students access to tools that aid in their learning.

- Textbook
- Lesson Videos
- Problem-Solving Help
- Submit a Problem Fix

> Note: Students will not have access to the Help tab during any exam. This includes the Pre-Test, Unit, Mid-Term, and Final Exams.

Each of these options is described in detail in the following sections.

Textbook

The Textbook (shown in Figure 4.7) contains written lessons that correspond to the video lessons in a course. Each Textbook lesson covers the same concept as the lesson video and is designed to be an additional instructional resource to enhance the learning experience.

Many Textbooks include the Main Concept of the lesson along with emphasis boxes that highlight important information students should know.

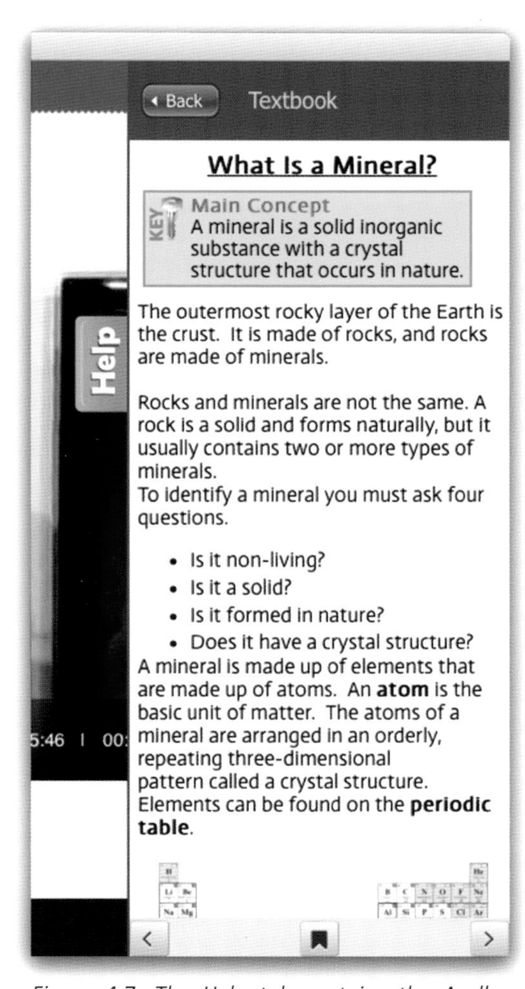

Figure 4.7: The Help tab contains the Acellus Textbook. The Textbook provides a written form of instruction about the concept the student is working on. The Textbook is not accessible during exams.

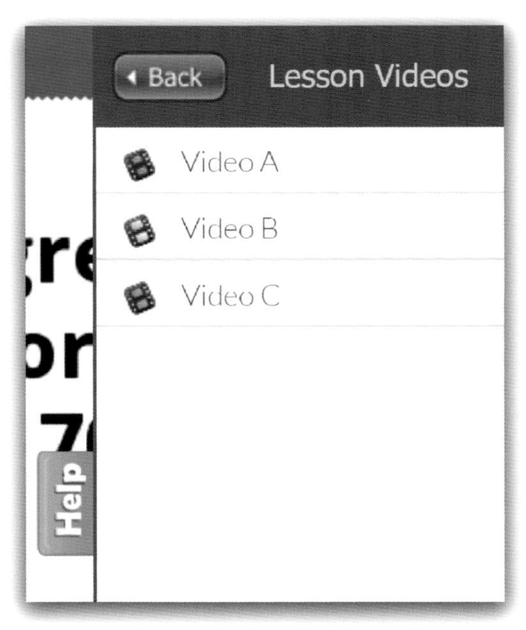

Figure 4.8: The Help tab provides access to the Lesson Videos. At any time during the practice problems or reviews, students can choose to watch any of the available videos filmed for the concept.

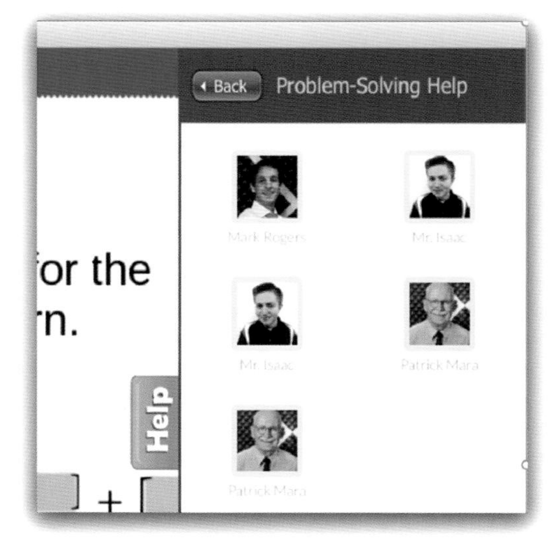

Figure 4.9: Problem-Solving Help Videos show students how to work a problem similar to the one they are currently struggling with. This short video is initiated by the students and gives support the moment it is needed.

Lesson Videos

The Help tab also includes the Lesson Videos option (Figure 4.8). Here students can access all of the available videos filmed for the concept. Sometimes seeing additional examples or hearing the topic explained in a new way makes all the difference. Students can view as many of these videos as they choose. When the selected lesson video ends, the students are taken back to the practice problems – right where they left off.

Problem-Solving Help

Help Videos are accessible through the Problem-Solving Help option under the Help tab. Here students have access to short, focused videos designed to help them learn how to solve a specific type of problem (Figure 4.9). In these videos, students see a sample problem being worked out with an audio explanation of what is being done and why. Through Help Videos, stuck students are able to reach out and get the needed assistance – right when they need it – to continue making forward progress in their studies.

Note: Problem-Solving Help Videos have been found to be a particularly helpful resource for students who are stuck on a problem and need a quick, focused review of how to solve it.

Submit a Problem Fix

Through the Submit a Problem Fix option, students are able to send a message to the Acellus Courseware Development team any time they believe they find something in a course that needs to be fixed.

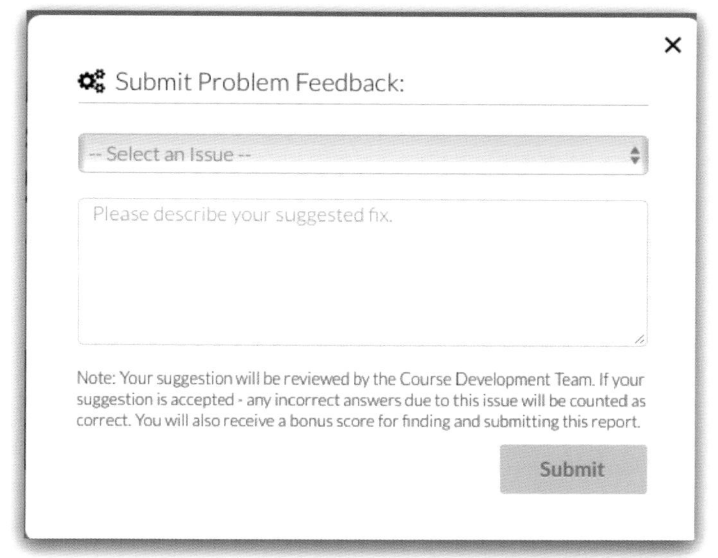

Figure 4.10: When students Submit a Problem Fix, they can choose the type of issue and describe their suggested fix.

On the Submit Problem Feedback screen shown in Figure 4.10, students can recommend fixes to be made to Acellus course content.

Each suggestion is reviewed by the Acellus Courseware Development team, and if the student's recommendation is valid, the suggestion is approved, the fix is implemented, and the student is given bonus credit for helping to improve Acellus.

Reviews and Exams

Acellus courses are divided into units, each including a unit review and exam. Midterm and final reviews and exams are also a part of each course. With an upcoming exam as a motivating factor, many students find the review to be a time when they dig in and work hard to make sure they have mastered the content on which they are about to be tested.

Although reviews have no direct impact on student grades, they provide students an additional opportunity to master course content before being tested on it. The reviews give them problems from each concept that will be covered in the exam. In addition, students have access to all of the Help features of Acellus. During the review students can read the textbook, re-watch videos, or get additional problem-solving help to ensure they understand and remember the content and are sufficiently prepared for the upcoming exam.

As soon as the review is complete, the Help features of Acellus are no longer accessible and the student begins the exam.

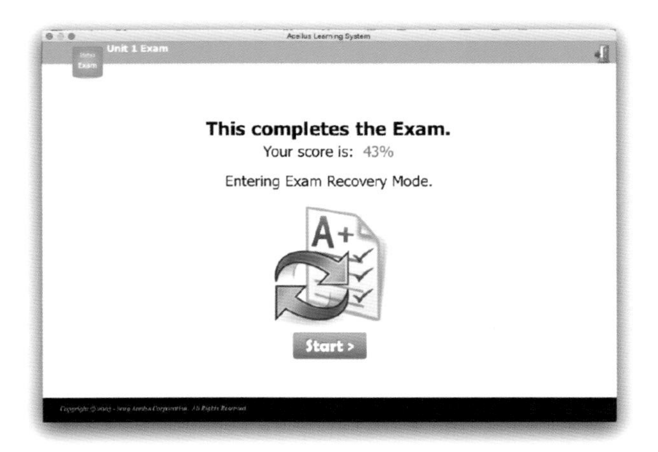

Figure 4.11: When students fail an exam, they are automatically placed into Exam Recovery Mode, where they can review the material they missed and retake the exam.

Exam Recovery Mode

Students passing an exam advance to the next section of their Acellus course. Those unable to pass an exam, however, are automatically placed in Exam Recover Mode (Figure 4.11).

Exam Recovery provides students with focused instruction on unmastered concepts. Here students see their wrong answer along with each problem they missed and are given another opportunity to solve the problem. If they still answer incorrectly, Acellus provides video instruction to reteach the concept. During Exam Recovery, students also have access to all of the Help features of Acellus.

Once students correctly answer all the problems they missed, they retake the exam with new problems that test them on the same concepts.

Retake for Extra Credit

Sometimes students pass an exam but want to earn a better grade. Acellus gives students this opportunity.

When students pass an exam with a less-than-perfect score, they receive a choice to either Continue, moving on to the next part of the course, or to Try Again for Extra Credit (see Figure 4.12). If students choose to try again, they see the problems they missed on the exam, including the answers they entered, and receive any needed additional instruction. Upon completing the review of missed concepts, these students take a new exam, getting another opportunity to be tested on these concepts.

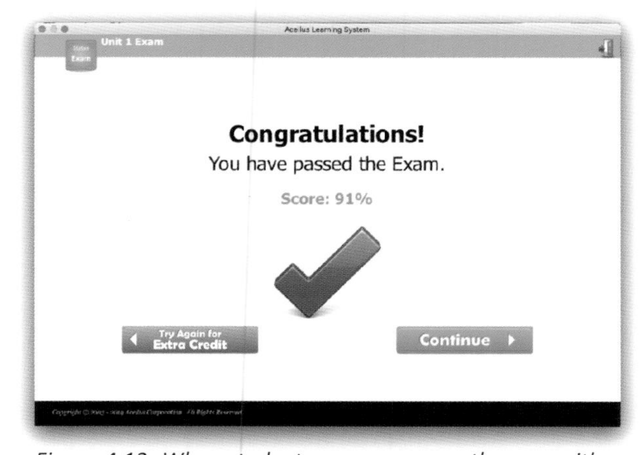

Figure 4.12: When students pass an exam, they can either Continue or Try Again for Extra Credit.

Vectored Instruction (Recovery Modes)

Very often, students lack a solid foundation of understanding or are missing important concepts needed for academic success.

Acellus uses Prism Diagnostics to identify when a student is struggling on a concept. If a chronic deficiency is detected, Vectored Instruction kicks in and the student is prompted that they are entering "Recovery Mode," as shown in Figure 4.13.

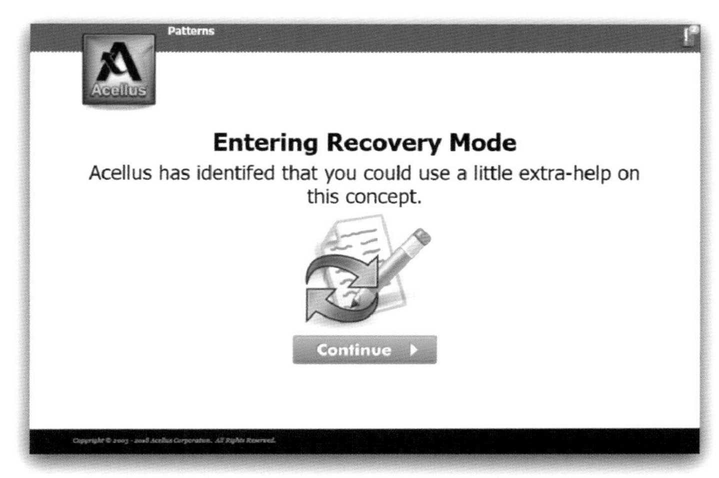

Figure 4.13: When students need help with more basic concepts, Acellus provides them with the necessary instruction.

In Recovery Mode, the student receives a series of lessons and problems taught on a more basic level. If students struggle during the first level of recovery, Acellus moves them into Deep Recovery – which provides a more basic and thorough review of needed concepts.

Some students still have trouble even after two levels of recovery. In these cases, Acellus switches into the Foundation Building Mode where it builds the missing concept from scratch. Once recovery is completed, students return to their former position in the course.

When students enter one of these Recovery Modes, their teacher is prompted on the Live Class Monitor (see Figure 4.14). Teachers can look at the student reports to view details on completed recovery steps.

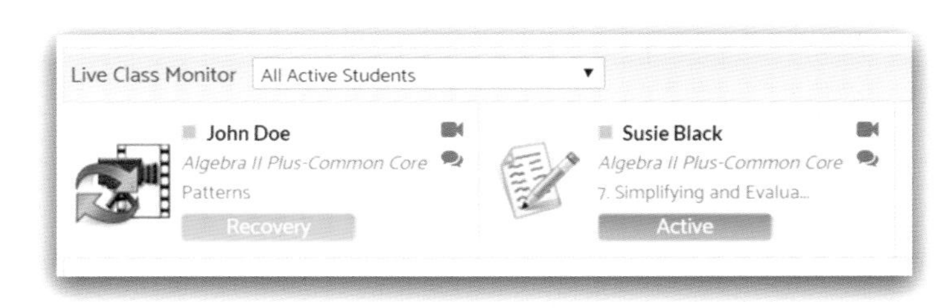

Figure 4.14: When a student enters Recovery Mode, the Live Class Monitor displays a special icon to give teachers a "heads up."

Memorization

In most classes, students are required to memorize small chunks of data. In general, the information that should be memorized is anything that is worth knowing that cannot be derived. Though necessary, memorizing can be tedious. Acellus takes the drudgery out of memorization using songs and drills.

Songs

One of the ways Acellus facilitates memorization is through the use of memorization songs. These music videos provide students with an engaging way to remember important information. An example includes the "Count to One Hundred" video shown in Figure 4.15.

Figure 4.15: Setting facts to music helps make them stick in students' minds.

Drills

Memorization Drills are another way Acellus helps students memorize (Figure 4.16). Some of the things these interactive memorization drills have been developed for include the following:

- Sight words
- Vocabulary
- Grammar
- Facts
- Spelling
- Math Facts

Recent development efforts at Acellus have focused on the use of cognitive science to improve memorization. For example, the math fact drill reinforces students' mastery of math facts by having them focus on pairing numbers that add up to a specific sum while requiring them to subconsciously pull math facts from their implicit memory.

Figure 4.16: From the top down, these images are examples of Sight Word Drills, Vocabulary Drills, and Math Fact Drills. Other types of drills include Grammar, Fact, Spelling, and Math Drills. Audio is often used in drills.

Additional Student Resources

Besides the online course instruction delivered by Acellus, students are provided additional resources to help them manage and enrich their learning experience. These resources are discussed in the following sections.

Progress

Under the Progress button, students can see how they are progressing in meeting their class goals (Figure 4.17). Here students can see their weekly goal along with a list of concepts they need to complete that day to be on schedule to achieve their class goal. Students can also see how far they have progressed through the course.

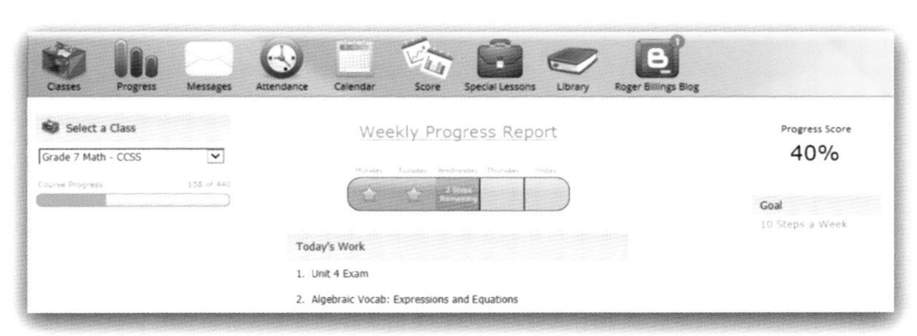

Figure 4.17: On the Progress screen, students receive feedback on how well they are progressing through the course.

Messaging

Students and teachers can interact with each other via the Acellus Messaging system, shown in Figure 4.18. This allows students to seek help, while giving teachers the opportunity to prioritize the needs of students, helping those first who need it most.

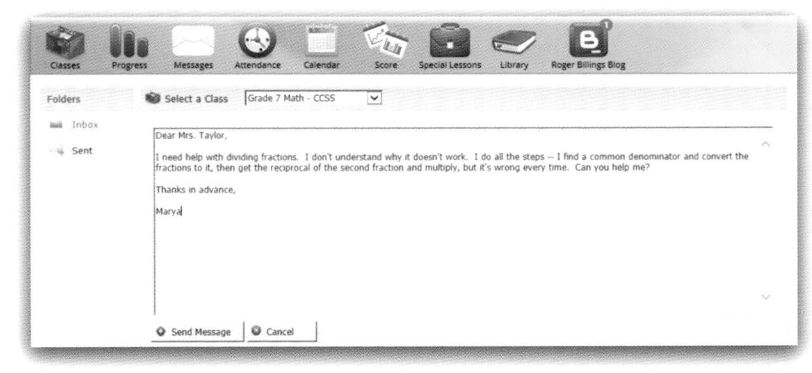

Figure 4.18: Students can send teachers a message when they need help, making it simpler for them to request help – and simpler for teachers to give it.

Attendance

The Attendance Report, shown in Figure 4.19, enables students to quickly review their attendance to make sure they are on track for meeting their learning goals and to identify courses where they need to spend more time.

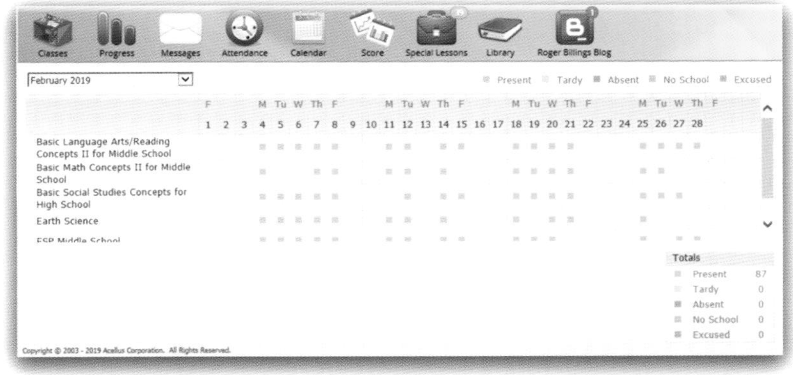

Figure 4.19: Students can see their own attendance record, color-coded for clarity.

Calendar

School events are posted on the Calendar. With the click of a button, students can view school events for the month, like those shown in Figure 4.20. They can also see previous and future school events.

Figure 4.20: Teachers can use the Calendar to make students aware of important upcoming events.

Scores

From the Score button, students can see their overall GPA and details on their performance in each of their courses (Figure 4.21). In addition to seeing their current grade in each course, students can see the number of Special Lessons yet to be completed and whether or not they have accomplished their progress goals.

Figure 4.21: Students can keep track of their scores and see where they need to improve.

By clicking on a specific course, students can see additional details, including their grade for each completed lesson and exam. Students wanting to review a prior concept or improve their grade can click on the Retake for Credit button to repeat an individual problem step.

Special Lessons

Special Lessons are supplemental learning resources activated by the teacher for use in the blended classroom. Special Lessons include activities such as writing assignments, classroom discussions, experiments, and projects.

Teacher-selected Special Lessons are automatically assigned to students as they progress through the course. When students have outstanding Special Lesson assignments, a small "bubble" appears on the briefcase icon on the student's menu, indicating the number of outstanding assignments (see Figure 4.22). Students select the briefcase icon to view their current Special Lesson assignments.

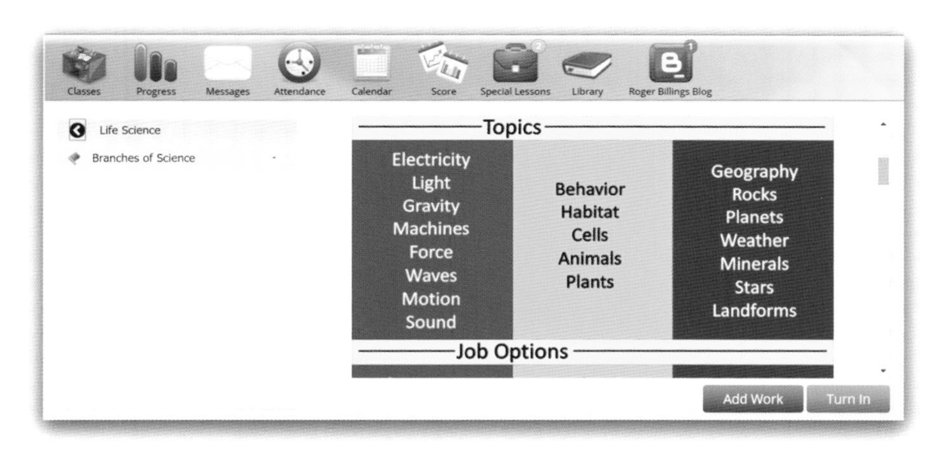

Figure 4.22: When a Special Lesson is assigned, a yellow bubble appears on the briefcase icon.

Newly assigned Special Lessons are marked with a blue flag. To submit their assignment, students select Add Work (shown in Figure 4.22) and either upload a picture or create a file using the Acellus text editor. After their work is submitted, the flag changes to green, indicating that they have turned in the assignment.

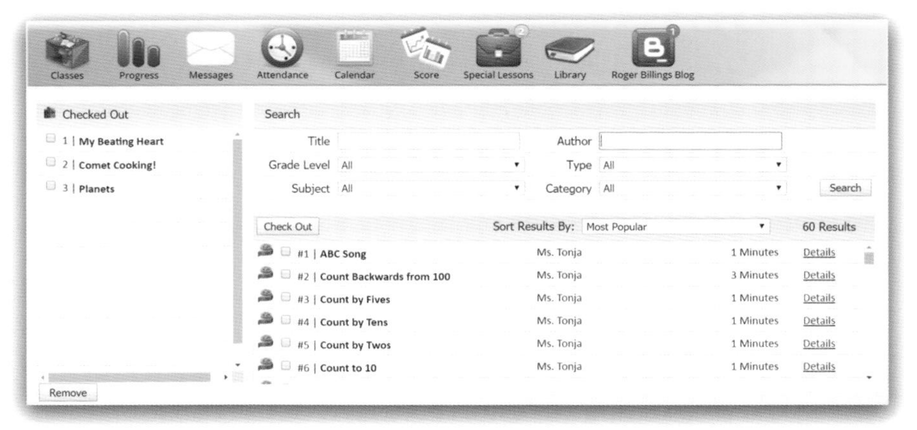

Figure 4.23: When either a student or a teacher checks out an item from the Library, it appears in the Checked Out list.

Library

The Acellus Library (Figure 4.23) offers additional educational material for classroom use. Teachers can check out media from the Library and assign it to their class. Students, too, can visit the Acellus Library and check out content to complete lesson assignments or to enrich their educational experience.

The Roger Billings Blog

The Roger Billings Blog for Students is a fun and informative blog designed to inspire Acellus students to achieve greatness (Figure 4.24). Short videos on interesting science topics, technological innovations, and inspirational and motivational stories help students get excited about science, their lives, their future careers – and begin looking for ways to make a positive impact on the world around them.

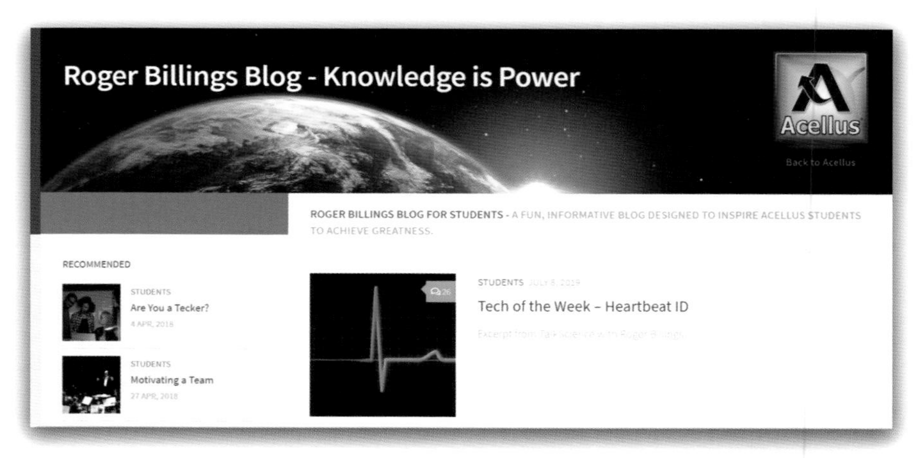

Figure 4.24: The Roger Billings Blog for Students presents new technologies and inspirational and motivational messages.

Parent Registration

In the upper-right corner of the student interface, students can select a link – Parent Registration – and enter the email address of up to three parents and one advisor (Figure 4.25). People registered here receive an email inviting them to monitor this student's progress from the Acellus Parent Interface. The Acellus Parent Interface uses GoldKey security to secure and protect student information.

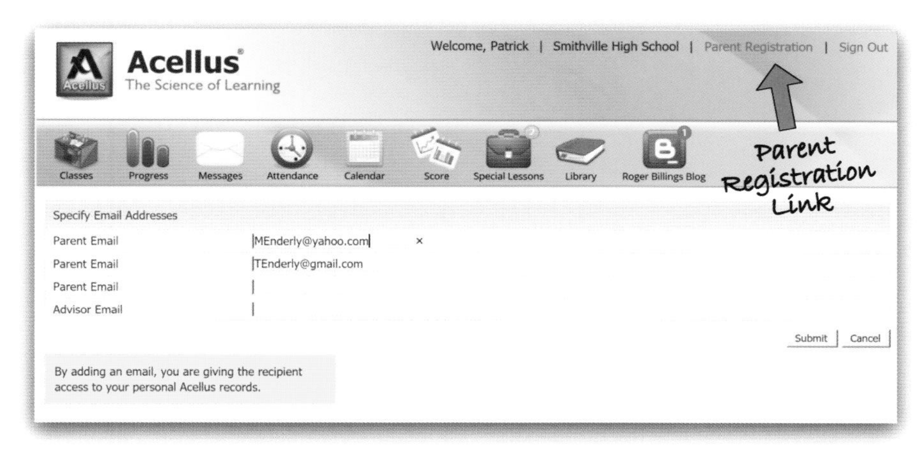

Figure 4.25: Students can insert a parent's email address using the Parent Registration link in the upper-right corner of their Acellus student interface. The parent will receive an invitation by email to monitor the student's work.

Individuals registered in this way can sign in to the Acellus App to access their student's records. Here parents can closely monitor their student's work on Acellus. In addition, at the beginning of each month, parents and advisors receive a detailed email report of the progress made by their student during the previous month.

Through these reports, parents and advisors can join forces with teachers to ensure that their student is getting the best education possible.

Chapter 5:

ACELLUS COURSEWARE DEVELOPMENT SYSTEM USER GUIDE

The Acellus Courseware Development System (ACDS, Figure 5.1) is a comprehensive, cutting-edge tool where Acellus courses are created, assembled, and enhanced. Teachers use the ACDS to view content as they work with students taking Acellus courses, while course developers use the System to create and refine course content.

In this chapter we will focus on the details of how teachers and course developers access and utilize the ACDS.

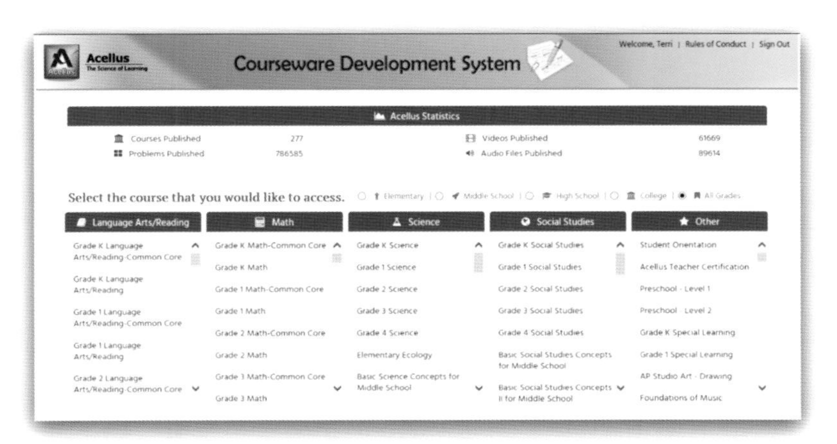

Figure 5.1: The Acellus Courseware Development System is a powerful tool that contains all the features necessary for teachers to view and developers to create Acellus course content.

Accessing the ACDS

Teachers and course developers who are registered in the Acellus System have access to the Acellus Courseware Development System through the Teacher Interface by selecting the Curriculum button on the gray menu bar, or alternatively going to http://courseware.acellus.com. The ACDS is protected by GoldKey security. Here you can select any course that has been published or is in the development process. For ease of searching, you can select Elementary, Middle School, High School, College, or All Grades to limit (or un-limit) which courses are displayed.

> Note: In the upper-right corner of the Courseware Development System's opening page you can select your name to edit your profile.

When you select a course, you will see the My Desk screen for that course (see Figure 5.2). There you will see tiles that link you to specific content areas such as those shown here.

Figure 5.2: The My Desk screen provides access to all of the elements of the course.

Course Content

The elements that make up Acellus courses include the following:

- The course syllabus
- Video lessons
- Problems
- Textbook
- Memorization drills
- Special lessons

The Syllabus

An Acellus Syllabus indicates the flow of the course – the content of the course listed chronologically. When you select the Syllabus tile (or the Syllabus button on the menu bar), you will see a screen like the one shown in Figure 5.3. The Syllabus gives you the following information:

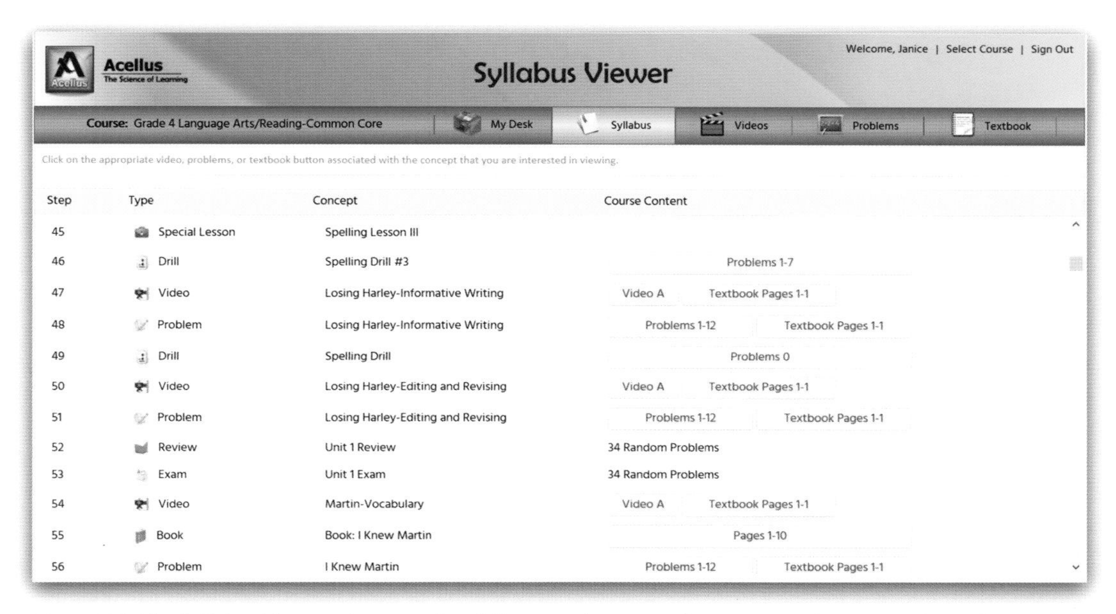

Figure 5.3: The Syllabus Viewer gives you the ability to review an entire course and to access the various course components.

- Step Number
- Step Type (Problem, Video, Textbook, Drill, Book, Special Lesson, Review, or Exam)
- Concept Name
- Course Content
  - Video A, B, C, D, as appropriate
  - Problems (with the current number of problems shown)
  - Textbook (with the current number of note pages shown)
  - Number of pages (for Books)
  - Number of problems in Reviews
  - Number of problems in Exams

The Video, Textbook, and Problems are accessible via buttons that you can select to see each of these course components.

ACDS for Teachers

Teachers helping Acellus students may find times when they want to view the actual course content. The Acellus Courseware Development System provides a way for teachers to view the videos and peruse the problems their students are encountering.

Viewing Videos

To review a lesson video, select the Video button from the menu bar or from My Desk, select the video button for the desired step, and the lesson video will start to play automatically.

Viewing Problems

Note: Special Lessons are listed on the Problems Page.

In the ACDS, teachers can view the lesson, drill, and book problems for any of the courses in the System, including both those that have been released and those that are under development. To do so, select the Problems button from the menu bar or from the My Desk screen. Next, select the Problems, Drill, or Book button for any step. You will see a screen similar to the one depicted in Figure 5.4, which includes a thumbnail list of the problems that have been created for that step. Things to be aware of:

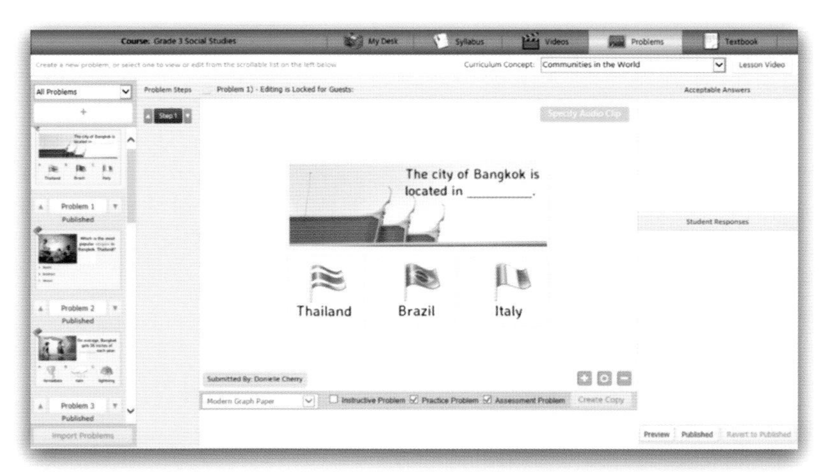

Figure 5.4: In the Problems section you can inspect all of the problems that have been created for the course. Select a Problem Thumbnail on the left to see that problem.

- By default, you will see a list of all problems in this step: both published and unpublished.
- By default, Problem 1 is shown in the active problem viewing screen.
- Select the Preview button (lower-right area of screen) to see the problem as it appears to your students.

Viewing Textbooks

The Acellus Textbook provides, in written form, the concepts covered in the video lessons. To view the Textbook for a given course in the ACDS, sign into the Courseware System, select the course, and then select the Textbook tile from the My Desk page, or select Textbook from the menu bar at the top of the page.

ACDS for Course Developers

The Acellus Courseware Development System is also the platform upon which course developers create lesson content and collaborate to study and refine the learning process. Based on sound scientific principles, Acellus is more than a tool for teaching: it also provides the feedback, data, and tools necessary for course developers to systematically develop and improve their courses and thus provide a progressively better learning experience for students.

The purpose of this section is to provide instructions for course developers who will be using the ACDS tools to create problems, Textbooks, and Special Lessons for an Acellus course.

Creating Problems

Creating problems within Acellus requires awareness of the principles that impact the problem content.

- Unlike traditional classrooms settings where teachers must often present multiple concepts in order to fill a lesson period, each video lesson in Acellus focuses on a single concept.
- The problems students see after each Acellus video lesson are specifically created for that lesson concept. Since problems are pulled randomly from a large pool, all problems for a particular lesson should test the concept taught in the lesson and should all be at the same level of difficulty.
- See "Writing Acellus Problems" in *Chapter 3: Developing Acellus Courseware* for more information about this topic.

Be sure to include appropriate instructions with each problem, whether as an audio file, as written instructions within the graphics or text element, or as implied instructions, as in the case of most text questions.

The basic steps for creating a problem include the following:

- Select Problems from the course page or from the menu bar of a course.
- Scroll to and select the lesson for which you will be writing problems.
- Open the problem template menu by selecting the "+" button located just above the thumbnail list.
- Select the template you want to use.

- Add the text, graphics, video content, and/or audio.
- Enter all forms of the correct answer.
- Set the problem type (see Figure 5.5 and accompanying description).
- Save the changes by selecting Post for Review above the main problem-viewing window.

Setting Problem Type

As you create problems, you should be sure to indicate whether each problem is for the accelerated (+), normal (0), or slower (-) mode of the course (see Figure 5.5). You should also indicate whether each problem is an Instructive Problem, a Practice Problem, or an Assessment Problem. This will tell the Acellus System how to use your problems to optimize the course for students.

Figure 5.5: The mode for the problem is set using the (+), (o), and (-) buttons near the bottom of the active problem viewer.

Choosing Problem Template

Acellus provides many different templates you can use to create problems. A general breakdown of the different problem types includes the following:

- Problems containing only text elements
- Problems containing only graphic elements
- Problems containing both graphic and text elements
- Problems containing scrolling text and text elements

- Problems containing a video element
- Custom "Blockly" problems for coding courses

Following is a detailed guide on how to create problems with graphic elements and problems with text elements, including scrolling text.

Adding Text to a Problem

We'll begin with the text elements. Let's say that we have started creating a new problem using a template that uses only text elements, such as the Text With 4 Long Text Buttons template. As its name implies, this template only requires text.

To enter the problem stem (or question), select the text that says "Enter Text Here" which is located where the stem should be – at the top of the problem area. A pop-up appears – a white area with a light-blue border – that also contains the words "Enter Text Here" (see Figure 5.6). In both places where this text can be seen, it is being pulled from the Acellus Database. This text must be removed or it will be saved back into the database again and will be visible to the students.

Figure 5.6: Select the area that you want to add text to. The Edit-Text pop-up displays the current text. In the lower-left corner of the pop-up you will see the current number of characters in the field and the total number of characters allowed.

Notice that in the lower-left corner of the border of the Edit Text field you see the number of characters currently in the stem – 15 at this point – as well as the number of characters allowed for this field – in this case 180.

Select all 15 characters and start typing to replace the default text. When you are finished, select Save. The pop-up will disappear and the text you have entered will appear in the main problem-viewing area.

Be aware that you will have different character limits for different fields and different templates as you will see when you select one of the answer fields where, for this problem template, you have up to 100 characters. You can edit what you've written (and saved) until you are happy with it. When you feel the problem is ready, select Post for Review. Your problem is saved and ready for review by the management team.

Adding Graphics to a Problem

Starting with the Image With 3 Image Buttons template, you will see a space for a larger graphic in the upper area of the problem where the stem will be and space for three smaller graphics below it, which will be the answer options.

Begin by selecting the large image area at the top to create the stem. The ACDS image editor opens (see Figure 5.7). Here you see the image editor where you will build the graphic and, to the right, the tools you will use to build it. Using the tools on the image toolbar, you can do the following:

> Note: The numbers on this list correspond to the red numbers on Figure 5.8.

1. Add text
2. Upload an image
3. Add an equation
4. Upload a GIF animation
5. Use the Clipart Library to find or upload an image
6. Duplicate an existing graphic element
7. Delete elements of the graphic

With the Size & Rotate tools and the Positioning tools, you can further affect the elements of the graphic.

Add Text to a Graphic

To create text in the graphic, select the text editor (No. 1 in Figure 5.8) and then click on the graphic area. A text box and a text menu bar will appear. Begin typing. You can change text attributes by dragging the mouse over the text you want to change and then using the text menu bar to format the font. One option for adjusting the wrapping is to adjust the width of the text box. To do this, make sure the text tool bar is not showing but the text is selected.

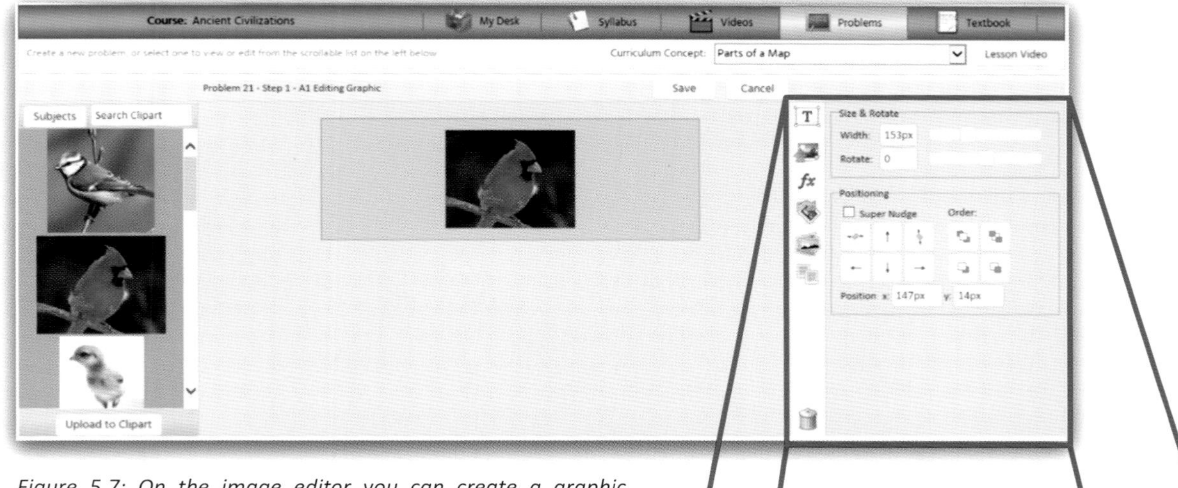

Figure 5.7: On the image editor you can create a graphic including images, text, and/or formulas/equations. The ACDS Clipart Library (shown on the left) gives you a large selection of clip art to choose from.

Upload an Image to a Graphic

On the image toolbar, tool No. 2 in Figure 5.8 allows you to upload clip art or an image that does not go into the ACDS clip art library. This means that it will not be available to other people who are creating problems in Acellus courses. When you select this button, you are given a file picker pop-up box. If you select a graphic, the image editor will remember it until you select a different one or sign out. That way, if you want to upload the graphic into this problem more than once, or into more than one problem, you can do so without having to search for it every time.

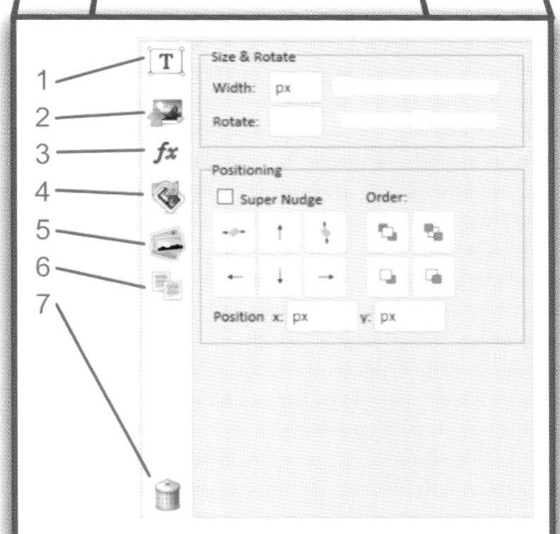

Figure 5.8: The tools in the image editor allow you to create text, import images, access clip art, and duplicate components. You can also size, rotate, move, and set the order of the components to create graphically pleasing problems.

Add an Equation to a Graphic

When you select the Equation button (tool No. 3 in Figure 5.8), the Equation Editor will open (see Figure

5.9). Here you can create an equation. You can type letters and numbers as desired, but you can also mouse over one of the menu items to see a fly-out menu of options, such as Greek letters. Select one to place it in the equation. When you have finished creating your equation, select Save Graphic. The Editor will close and your equation will appear in the problem area. Drag to place it where you want it. Select Save when you are finished.

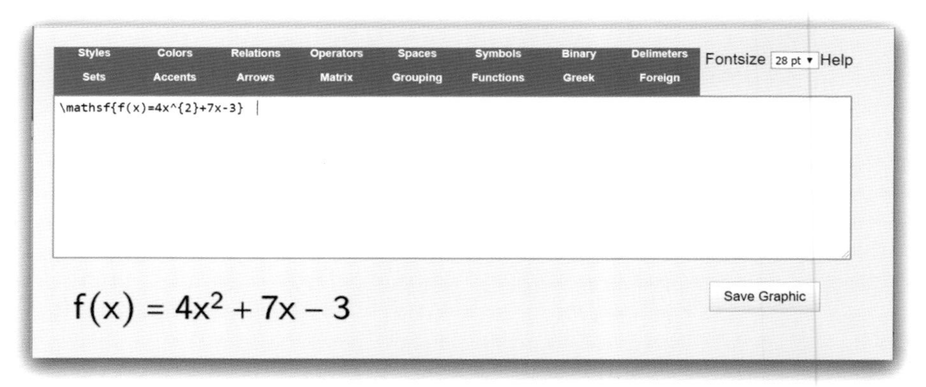

Figure 5.9: The Equation Editor allows you to create complex equations and insert them into problem graphics in the form of a graphic image.

Upload Animated GIFs to a Graphic

There is a special button for uploading animated GIF files – No. 4 in Figure 5.8. This button behaves much as the Upload a Graphic button (No. 2 in Figure 5.8). Files uploaded this way are not available to other problem creators.

Use the ACDS Clipart Library to Find or Upload an Image

You can also upload a graphic to the ACDS Clipart Library and then place that graphic in the problem. This makes the graphic also available to other people who are creating problems for Acellus courses.

Selecting the Open Clipart tool (No. 5 in Figure 5.8) displays the Clipart Library on the left side of the screen. To upload a graphic to the Clipart Library, select the Upload to Clipart button at the bottom of the Clipart Library. Enter the word(s) that best describe the subject of the clip art, such as “Girl” and then enter a brief description, such as “Young blonde girl holding a cat.” Choose the file and upload it; then find it under the subject you provided. Select it to add it to your image. You can also use the Size & Rotate tools and the Position tools to adjust these images.

Duplicate Graphic Components

The tool that looks like two identical pieces of paper (No. 6 in Figure 5.8) is the duplicator tool. Select any component of the graphic you are building and click this button. A second copy of the component will appear, offset a little and in front of the original. You can use the Size & Rotate and Positioning tools to place this new component right where you want it.

Scrolling Text Problems

In some courses students need to be able to read a text – a story, a play, an essay, etc. – or view a graphic – a map, a graph, a formula, etc. in order to answer the question appropriately. For these situations, Acellus provides Scrolling Text templates (Figure 5.10). Here the problem creator can insert text or graphics or both into the scrolling text field, which will appear alongside the question and answer options.

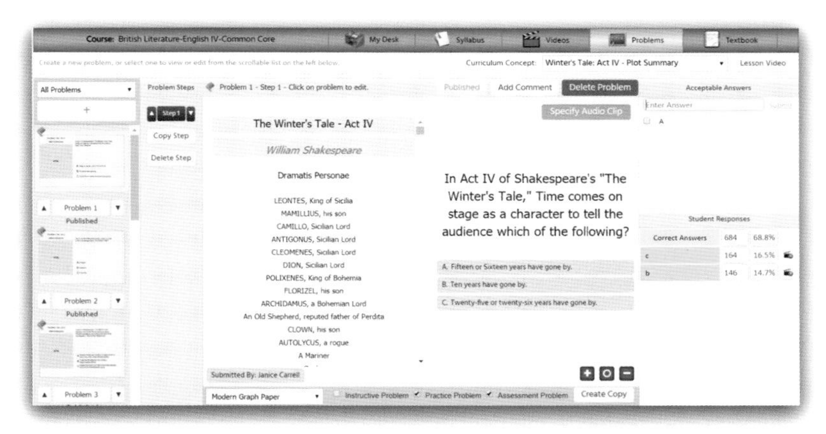

Figure 5.10: The Scrolling Text with Text Buttons template provides a way to include texts and graphics that students need in order to answer questions properly.

When you click on "Enter Text Here" below the titles, the Scrolling Text editor opens (see Figure 5.11). Here you can enter and format text, insert graphics, and access the ACDS Clipart Library. The tools are the same as those used to create a Textbook, and are described in detail in the section titled Creating the Textbook below.

> Note: Be sure to respect copyright laws when creating Acellus course content.

When you are finished, select Save, and then be sure to preview the problem. You can edit anything you're not happy with.

Above the scrolling text field are two additional fields that you can use for title, author's name, or other appropriate information. These fields are used just like the text fields described above.

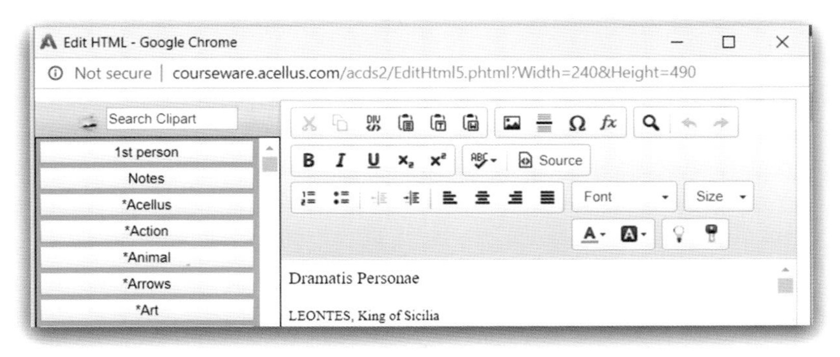

Figure 5.11: Select the Scrolling Text Field to edit its contents here.

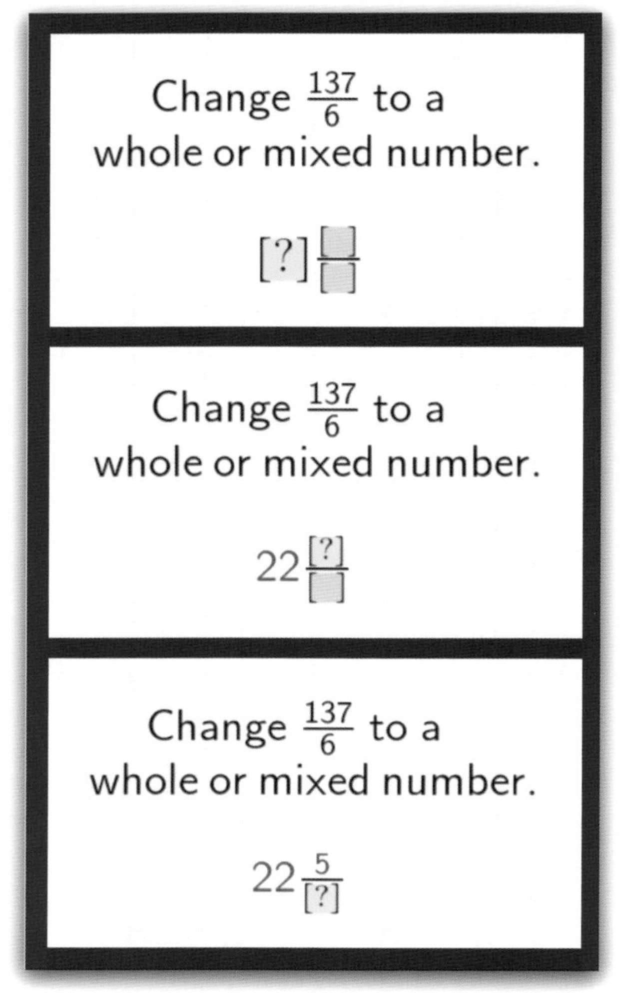

Figure 5.12: Step problems provide a way for students to enter multiple-answer questions in a straightforward and easily understandable format. Students can focus on one step at a time, making the problem manageable. This can also help Acellus to diagnose the exact knowledge bit that students aren't getting.

Step Problems

A Step Problem is a problem where the student enters the answer in steps. A simple example is the problem shown in Figure 5.12 whose answer is twenty-two and five-sixths. The answer could be entered "22 5/6," or the answer could be broken into steps. If entered by steps, the students will see a green box where the 22 should be, and gray boxes where the 5 and the 6 should be. The students are instructed to enter the number that goes in the green box.

After they enter the 22 correctly, they see the same problem again with the 22 in its place, a green box where the 5 goes, and a gray box where the 6 goes.

After they enter the 5 correctly, they see the problem with both the 22 and the 5 in place, and a green box where the 6 goes. They enter the 6 and move on to the next problem.

Any problem template can become a Step Problem. To create a Step Problem, add a problem by selecting the "+" button and choosing one of the problem templates, and then select Copy Step, located between the thumbnail list and the active problem viewing screen (see Figure 5.13). The new copy is automatically displayed in the active problem viewing screen. Edit the problem as needed. If you need another step, repeat this process. Be sure to enter the correct answer for each step.

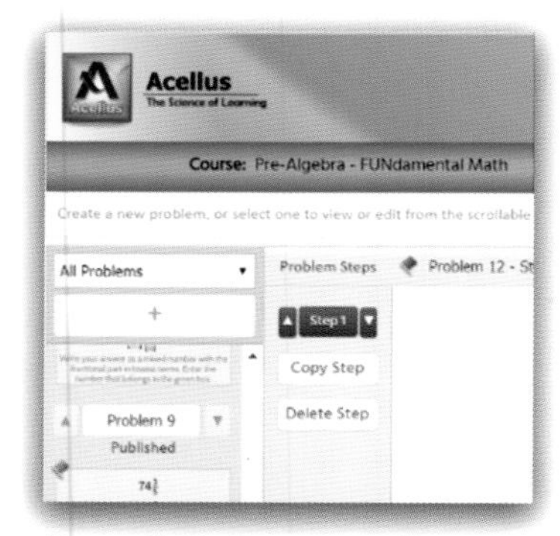

Figure 5.13: You can easily create a step problem using the Copy Step button located between the thumbnail list and the main problem view screen. When a problem is published, the Copy Step option is no longer visible.

Creating the Textbook

The Textbook is a written companion to the video lesson that is designed to do the following:

- Provide a supplementary method of presenting the lesson concept
- Strengthen students' understanding of the lesson concept
- Help students move the concept into long-term memory
- Help students learn to take effective notes

To access a course Textbook within the ACDS, select Textbook on the menu bar or from the My Desk screen. This will automatically bring up the Current Version of the Textbook, as indicated by the tab on the left.

Since the Current Version cannot be edited, to make changes in the Textbook, select the Create New Revision button in the upper right corner of the Textbook editor (shown in Figure 5.14). When you have made the desired changes, select Post to have your work reviewed for approval.

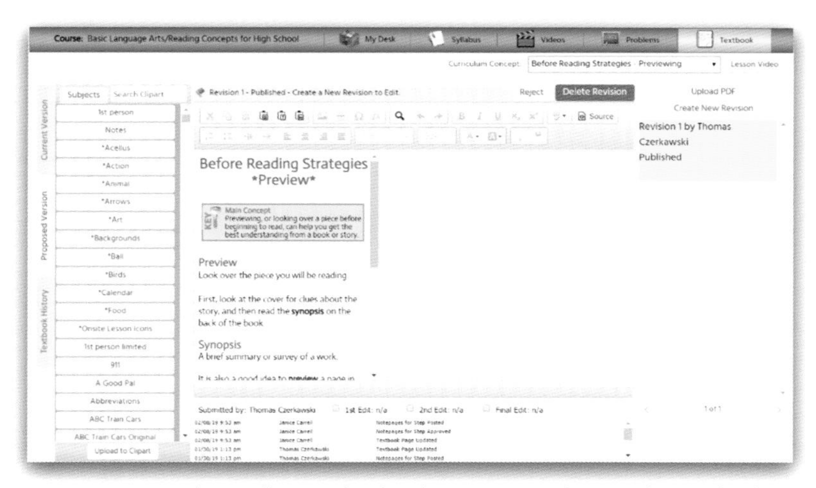

Figure 5.14: In the Acellus Textbook editor, you can format the Textbook in HTML, or you can upload a PDF file.

When you select Textbook from the menu bar or from the My Desk screen, you will be able to see in-process versions of the Textbook, which make no impact on the currently-published version.

The Textbook Editor is flexible and powerful, with two rows of tools for adding and editing text and graphics, as well as for creating equations (Figure 5.15).

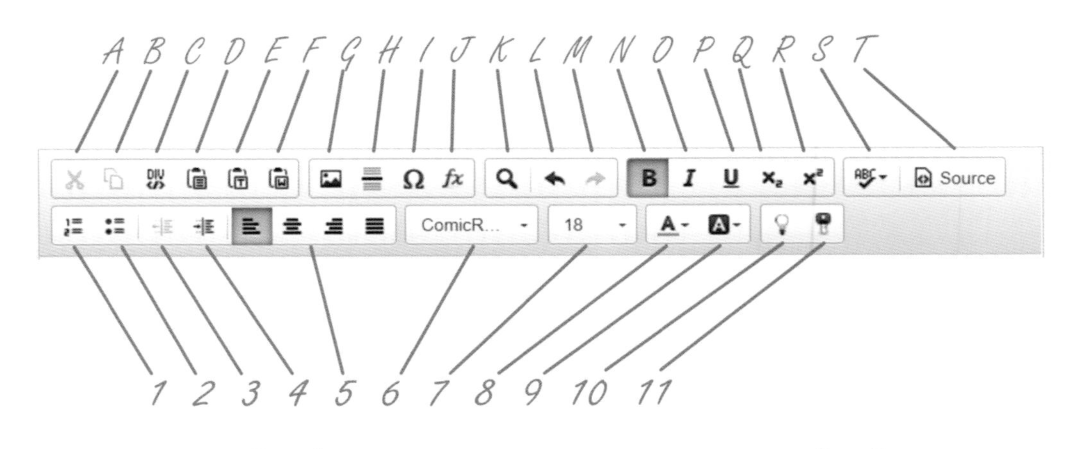

Row 1

A	Cut
B	Copy
C	Create a DIV Container
D	Paste
E	Paste as Plain Text
F	Paste from Word
G	Upload a Picture (not to Library)
H	Insert a Horizontal Line
I	Insert a Special Character
J	Edit an Equation
K	Find
L	Undo
M	Redo
N	Bold
O	Italic
P	Underline
Q	Subscript
R	Superscript
S	Turn on/off Spell Check while typing
T	Edit Source Code

Row 2

1	Insert / Remove Numbered List
2	Insert / Remove Bulleted List
3	Decrease Indent
4	Increase Indent
5	Align: Left/Center/Right/Full Justify
6	Font Name
7	Font Size
8	Text Color
9	Background Color
10	Insert Hint
11	Insert Key (used for main idea)

Figure 5.15: A full set of tools for editing HTML makes the Textbook Editor very powerful. The red letters and numbers, above, correspond to the letters and numbers in the table.

Note that if you are doing the writing in Microsoft Word, you should use the special "paste from Word" button (F in Figure 5.15). Text from this particular word processor requires special handling for importing.

There is a preformatted template for displaying the Main Concept of the lesson (the KEY icon – see No. 11 in Figure 5.15), and another for displaying a HINT – important points from the lesson that students should remember (the light bulb icon – see No. 10 in Figure 5.15).

The Upload PDF button is located on the right side of the screen, above and to the right of the toolbar. Use this button to browse for and select the correct PDF file and upload it.

Above the Upload PDF button is the Lesson Video button. You can watch the video without leaving this page. The video comes up in a new window, which you can drag around the screen so that you can review your notes while it is playing. The video includes full video controls to allow skipping forward and back for efficient editing.

When you have completed all your changes to the Textbook revision, the management team will review your work for any additional edits it needs. Once it has been approved, the revision is published. The previous version will then be listed in the Textbook History, which is also accessible via a tab on the left.

Creating Special Lessons

To create Special Lessons you will use a procedure much like that for creating Problems. Open the Problems section, then locate the Special Lesson step that you want to create. Select the Problem 0 button for that step. You will see an area labeled Special Lesson Resources, with a place for the Student Resource and a place for the Teacher Resource (see Figure 5.16).

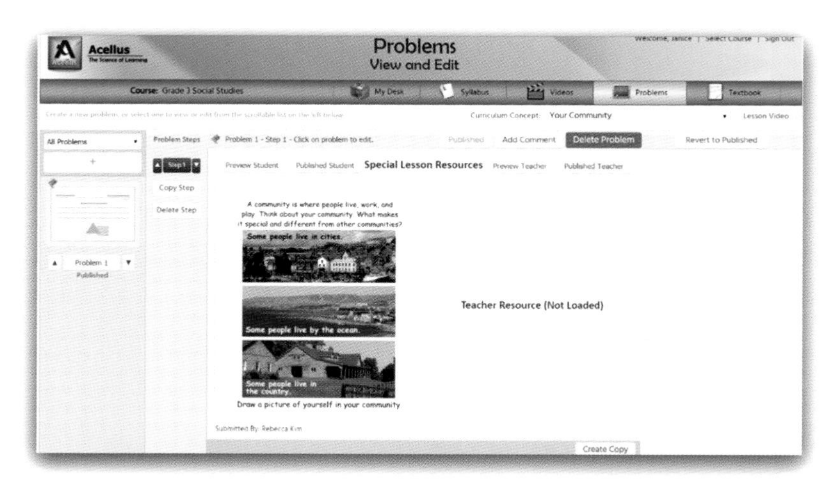

Figure 5.16: In the Problems – Special Lessons section of the ACDS you can select a Special Lesson to view or edit. Here you can see both the Student and the Teacher Resource for the Special Lesson. Click one of these to edit it.

For each, you have the option of either creating a graphic in much the same way as you would create a Problem graphic, or uploading a PDF file. Whichever you want to use, click on the Resource area you want to edit (Student or Teacher) to enter the editing mode. As shown in Figure 5.17, in the editing area for the Special Lesson resource you will see the same tools as those found in the Problem-editing area, as well as a tool to upload a PDF. Once you have created the Resources, you can preview them and edit them as desired. When the Resources are ready, post them for review.

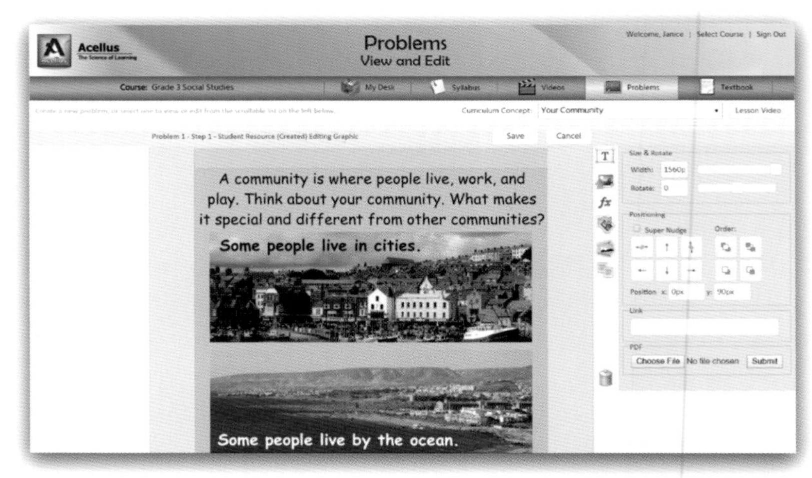

Figure 5.17: The editor for Special Lessons is very similar to the editor for problem graphics. Special Lesson creators have the option of editing the graphic this way or uploading a PDF file – notice the button in the lower right corner of the editor screen.

Summary

The Acellus Courseware Development System is a very powerful technology that allows course developers to collaborate in creating effective, customized course content. This system is also useful for teachers who want to review course curricula – including the problems or videos that their students are seeing – but, further, who want to join the Acellus Courseware Development Team and use their expertise to improve these courses, making a critical, positive difference in the lives of millions of students every day.

APPENDIX

Contact Information

If you have any questions about using Acellus or GoldKey, or you have problems that are not covered in this book, please contact the International Academy of Science:

Contact Customer Service

Via phone: 877-411-1138
Via email: customersupport@acellus.com

Customer Service Hours:

8:00 am to 5:00 pm Central Time, Monday through Friday

Please feel free to leave a message or email us after hours, and we'll get back to you as soon as possible on the next business day.

Trademarks

"Acellus" and "Prism Diagnostics" are registered trademarks of Acellus Corporation.

"Vectored Instruction" and "Learning Accelerator" are trademarks of Acellus Corporation.

"GoldKey" is a registered trademark of CybrSecurity Corporation.